Water on the Great Plains

Issues and Policies

Water on the Great Plains

Issues and Policies

Peter J. Longo and
David W. Yoskowitz, Editors

Texas Tech University Press

Library of Congress Cataloging-in-Publication Data
    Water on the Great Plains : issues and policies / Peter J. Longo and David W. Yoskowitz, editors.
        p. cm.
    Includes index.
    ISBN 0-89672-459-X (cloth : alk. paper)
    1. Water-supply—Great Plains. 2. Water-supply—Government policy—United States. I. Longo, Peter Joseph. II. Yoskowitz, David W.
    TD223 .W29197 2002
    333.91'00978--dc21
                                          2001001832

02 03 04 05 06 07 08 09 01 / 9 8 7 6 5 4 3 2 1

Texas Tech University Press
Box 41037
Lubbock, Texas 79409-1037 USA

1-800-832-4042
ttup@ttu.edu
www.ttup.ttu.edu

This is dedicated to Charlie, Amy, Diane, Sarah, and Peter Longo, and to Bob Miewald.
P. J. L.

To Carolyn, my family, and Jim Jonish, who put me on the path (river) of water research.
D. W. Y.

Contents

Foreword

Much has been written about the Great Plains, once termed the "Great American Desert" by early explorers. Contrary to those early first glimpses, all the ingredients of excellent agricultural productivity mark the Plains and its hardy inhabitants. The climate, the soils, the rainfall (although sometimes short), the availability of groundwater and the citizens all contribute to the shared culture of the region. This excellent book, co-authored by Peter Longo and David Yoskowitz, lends wonderful insight to a myriad of water issues that must be and are being considered in the Great Plains. Just like the climate, other things tend to change rapidly in the area. Municipal, industrial, and wildlife demands have greatly increased in recent years. While serving as the Governor of Nebraska, I was very involved in the effort to develop a Memorandum of Agreement that led to the three-state Cooperative Agreement between Colorado, Wyoming, and Nebraska and signed by each governor. The Federal Endangered Species Act was the driving force to begin settling the longstanding conflict over endangered species needs and water diversion from the Platte River.

In Chapter I, Charles Bicak points out, "Although it is unclear how the future of the Great Plains will unfold, there is no question that water issues will be at the center of decision-making." Decision-making may even need to be accelerated, as water is an essential ingredient wherever you go, and it is certainly one of the most essential in the Great Plains. In Chapter III, John Anderson describes conflict avoidance and describes it as the kind of medicine most people prefer. How to administer the medicine is another problem; to some, solving critical water issues maybe somewhat like taking castor oil. However, every effort must be made to resolve conflicts. That effort has been in full swing on the Platte River and is described by J. David Aiken in Chapter IV. Peter J. Longo makes an excellent point in Chapter V regarding the economic drain caused by water conflicts. I can vouch for that with firsthand experience of water conflicts and lawsuits while serving as Governor of Nebraska. Brian Ellison does well in Chapter VI developing the case that the development of natural resources is now an intergovernmental task. We are very aware of the fact that water development will not occur unless agencies work to address diverse interests and goals.

While I have not commented on all of the issues and ideas brought out in *Water on the Great Plains: Issues and Policies,* the economy of the Great Plains is tied heavily to agriculture and the importance of water will continue to be a key ingredient for sustainability. Conflicts must be resolved in the most efficient way possible and teamwork is essential.

E. Benjamin Nelson
United States Senator
Nebraska

Acknowledgments

There are many people who helped make this book a reality. The steadfast assistance provided by Texas Tech University Press was beyond the call of duty.

The idea percolated from a summer Peter Longo spent at the University of Manitoba with the support from a grant from the Government of Canada. Special thanks is directed to the faculty and staff of the University of Manitoba Natural Resources Institute, particularly Professor Thomas Henley. To push this project, Peter received generous sabbatical support from the University of Nebraska at Kearney, Research Services Council, and the University of Nebraska Water Center, especially Robert Kuzelka, Bob Volk and Marion O'Leary. Additional sabbatical support was provided by the Department of Political Science at the University of Nebraska-Lincoln. The late Professor Robert D. Miewald served as the best mentor and advisor. Peter is grateful for the constant support from his colleagues in the department and campus. Special thanks go to research assistants Stephanie Wileage, Susan Champlin, Clayton Thyne, and Lucas Swartzendruber, and to secretary Barb Harshbarger. Finally, Peter wishes to thank his always positive and loving parents, Charlie and Amy, his loving and supportive wife Diane and children Sarah and Peter, the greatest siblings (eight older brothers and sisters), Fr. John Schlegel, a sibling-like professor, and his understanding in-laws Janet and Gerry Haney.

David Yoskowitz became involved in this project through his affiliation with the Association of Arid Lands Studies and would like to thank them. To help complete the project, David received summer grant monies through the College of Business Administration at Texas A&M International University. David would like to thank David Hudgins and Mike Pisani in their roles as sounding boards. The strong support that David receives from his family has always served him well. He thanks Bill, Renee, Peggy, Issie, Michele, Fred, Laurie, Peggy B, the Grandparents, and Uncle B.

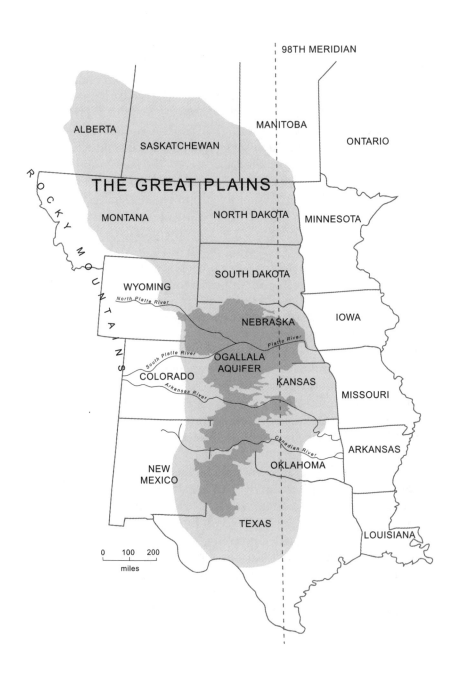

98TH MERIDIAN

ALBERTA

SASKATCHEWAN

MANITOBA

ONTARIO

THE GREAT PLAINS

MONTANA

NORTH DAKOTA

MINNESOTA

SOUTH DAKOTA

WYOMING

North Platte River

NEBRASKA

IOWA

South Platte River

Platte River

OGALLALA
AQUIFER

COLORADO

Arkansas River

KANSAS

MISSOURI

Canadian River

ARKANSAS

NEW
MEXICO

OKLAHOMA

LOUISIANA

TEXAS

ROCKY MOUNTAINS

0   100   200
miles

Introduction

Peter J. Longo and David W. Yoskowitz

The environment of the Great Plains of North America has stirred the passions and aroused the will to survive of many an inhabitant. From Texas to Alberta, residents of the plains are bound by a shared appreciation of the region. The casual onlooker undoubtedly may not understand this shared appreciation found on the plains. Still for others, like Ian Frazier (1989), a strange appreciation can hit the outsider:

> For hours we drove on roads the Rand McNally & Company considers unworthy of notice. A moth glanced off the edge of the windshield, and in the sunset the dust its wings left sparkled like mascara. That night my friend said on a gas station pay phone, "I'm on the Great Plains! It's amazing here! The sky is like a person yawned and never stopped!" (15)

But to the careful and patient observer, the Great Plains is more than a yawn producer.

For the student of the North American Plains there is much to be found in the reality and perceptions of this vast region. Worster (1992: 105) states it best regarding the further academic work needed to be done on this region: "The ecological history of the Great Plains is still to be accomplished, still to find its historians. When they come to write it, they will have a subject of international significance, for these days the dry lands of the earth are under pressure and scrutiny." Worster is correct and we share in the call for research into the ecology of the Great Plains. Indeed, it is the ecology of the region that serves as a common denominator.

After all, Mother Nature—not some mystical shared culture phenomenon—imposes a connection for citizens from Texas to Alberta. Conventional wisdom suggests that residents of New England, the Deep South, and Quebec share the same culture. However, the culture of the Great Plains is not so conveniently viewed. Certainly, these inhabitants have a shared spirit to thrive on the plains, but the rich ethnic diversity of each state and province could discount any weight previously placed on the balance of this common ground. Nonetheless, there appears to be a sense that there is indeed commonality in the residents on the Great Plains of North America. Rather than strain to ascertain some mystical shared political culture, it makes greater sense to focus on the actual laws and issues relevant to the bi-national region and to the realities of nature. The laws and policies are the forces that bind these citizens; most important are the issues of the preservation of water. Water issues and the resulting policy and law are the binding ingredient to the citizens and political actors of the Great Plains.

With water and its scarcity being one of the more important issues in all the Plains States and Provinces, it becomes crucial to understand the role governments play to protect this most critical resource. Much is to be gained by comparing the policies that the regimes of the North American Plains produce. Comparative public policy, according to Heidenheimer, Heclo, and Adams (1990), is crucial for the following three reasons:

> One aim is to look for guidance in designing better policies. . . . A second aim in comparing public policies is to gain a deeper understanding of how government institutions and processes operate as they deal with concrete problems. . . . The third reason for studying policies across boundaries lies in the growing interdependence that is the hallmark of our times. (1–2)

However, where governments take over nature begins. The physical makeup of water and its importance to the biological-ecological system of the Great Plains is the focus of chapter 1 by Charles J. Bicak. He says, "Water is a remarkable, almost magical substance," and with this statement shows us how we can be affected by water at many different levels. It is the lack of water at various times throughout the period of European-American settlement that Conrad T. Moore discusses in chapter 2. His detailed investigation into the challenges faced by the early settlers as they tried to cope with severe drought is a wonderful lesson in history and geography. These are dimensions of water beyond the physical. What of the idea of a shared culture for the region? John Anderson, in chapter 3, leads us through the complications that arise when "culture" is used as a salve to help prevent the problems that water battles bring to a community and a region.

As water becomes scarcer and its relative importance grows, issues are often settled through a legal or political process. In chapter 4, J. David Aiken argues that although traditionally laws and institutions have been created to enhance the efficient use of water, this rarely leads to an equitable use and that equity should be a factor in the resolution of water disputes. Brian A. Ellison reveals the political side of water in chapter 4 where he examines the development of water resources by Denver, Colorado, and especially the controversial Two Forks water project of the late 1980s. No longer is it possible for the old water establishment (water commissioners, Army Corps of Engineers, etc.) to form an alliance and dictate water development. Because of the limited number of water resources and many shared bodies of water, conflicts arise between states and provinces. Peter J. Longo, in chapter 5, contends that "water conflicts are a drain on economic as well as natural resources and the conflicts tend to bring the worse out of parochial actors."

The economic aspects of water find its way into many arenas. Charles R. Britton and Richard K. Ford in chapter 7 examine Great Plains states' participation rates in various federal water grants to find that they are not homogenous in level of participation. In chapter 8, David W. Yoskowitz analyzes water-marketing activities on the West Texas Plains. The major cities of this region are actively acquiring groundwater rights to bolster their current supplies and plan for the future.

Not all of the action surrounding water takes place in the areas of law, politics, and economics. Cultural issues between seemingly differing regions creates an interconnectedness especially when it comes to the issue of water. This is evident in chapter 9 by Steven Danver who argues that water to the Pueblo peoples is as much an identity as it is a means of growing their crops. Joan M. Blauwkamp, in chapter 10, takes us into the realm of filmmaking and shows us the use of water as a central player in conservation politics. But as she states, "Local communities are best positioned as conservators of their home ecologies, including water. . . ." Four poems by Charles Fort tie together all the discussed issues and provide yet another means in which to express the power of water on the Great Plains.

REFERENCES

Frazier, Ian. 1989. *Great Plains*. New York: Farrar, Straus & Giroux.

Heidenheimer, A. J., H. Heclo, and C. T. Adams. 1990. *Comparative Public Policy: The Politics of Social Change in America, Europe, and Japan*. 3rd ed. New York: St. Martin's Press.

Worster, Donald. 1992. *Under Western Skies*. Oxford: Oxford University Press.

Part One

Physical and Cultural Dimensions

The Ghost of the World on the Plains

There was a ghost	*a white buffalo*
On prairie grass	*shot through its eye*
There was the eagle feather	*and angel wing*
Hooves and lightning	*on the plains*
Whose hands unearthed	*the charred bones*
The eagle feather	*the estuary of the heart?*
On prairie grass	*there was a ghost*
Shot through its eye	*on the plains*
The charred bones	*in a child's hand*
There was the eagle feather	*on prairie grass*
Shot through its eye	*a white buffalo*
There was the angel wing	*and eagle feather*
The estuary of the heart	*hooves and lightning*
The ghost of the world	*on the plains*

Charles Fort

The Rigors of Existence
on the Great Plains:
The Role of Water

Charles J. Bicak

Water is the limiting resource for most terrestrial plant growth and development west of the Mississippi River. Approximately two-thirds of the United States, including the Great Plains, receives less than thirty-five inches of precipitation a year. The Great Plains—from the Prairie Provinces of Canada to Texas in the south and from the Missouri River to the Front Range of the Rocky Mountains in the west—has an intense interest in the fate of water. Drought-tolerant grasses and shrubs predominate and successful agriculture relies upon regular irrigation, making the availability of water the key regulator of both native-plant communities and profitable crop production.

This chapter describes the properties of water, the availability of water in the Great Plains, current patterns of water allocation, current attitudes toward a water-use ethic, and some considerations for water availability in the future.

PROPERTIES OF WATER

Water is a remarkable, almost magical substance. Most of the time, humans tend to focus on issues of water quality and quantity: How clean is it? Is there too little? Too much? Often overlooked are the amazing things water does to and for living systems. Water is a liquid across the wide range of tolerance of living organisms: 0–100°C. Most other substances as light as water (H_2O: molecular weight = 18) are gases; hydrogen (H: atomic weight = 1.0), helium (He: atomic weight = 4.0), nitrogen (N_2: diatomic weight = 28), carbon dioxide (molecular weight = 44), inert gases such as neon (Ne: atomic weight = 20) and argon (Ar: atomic weight = 40), and halogens such as fluorine (F: atomic weight = 19) and chlorine (Cl: atomic weight = 35). The diffusive properties of these gases cause them to be ineffective in nutrient, oxygen, and energy transport in plants and animals. Water, by contrast, is the ultimate transporter. The liquid nature of water causes it to be highly effective in plant and animal systems. In plants, the pathway of water transport is from the soil to the root, to the stem and shoot, to leaves, and ultimately to the atmosphere via evaporation and transpiration (Figure 1). The connector along this soil-plant-atmosphere continuum (SPAC) is the nearly incompressible liquid nature of water. Water is an equally important hydraulic fluid in animals as it maintains the circulatory system. Without water as the carrier for hemoglobin, and hence oxygen, animal metabolism would not exist as we currently know it.

Water is "sticky." That is, molecules of water cling to one another (cohesion) and to other surfaces (adhesion) as well. Hydrogen bonds form between adjacent water molecules and among clusters of water molecules. These bonds tie water molecules together in a latticelike structure that allows the water molecules to "stick" to other surfaces such as the internal lining of the xylem conductive tissue in vascular plants. Water movement along the SPAC then is no small feat for plants. In anthropocentric terms, an ironic twist is that although the hydraulic nature of water and its hydrogen bonding ensure movement throughout the plant, loss of water via transpiration from leaves presents a very real risk of desiccation in nearly all months of the April-to-October growing season in the Great Plains.

Water may occur in three states: liquid, solid (ice), and gas (vapor). The unusually high amount of energy required to convert one gram of water to one gram of water vapor at 20°C (586 calories) is called the latent heat of vaporization. Similarly, the amount of heat required to melt ice is relatively high—eighty calories at 0°C. This has implications for plants in several ways. Although water loss by transpiration places Great Plains' vegetation at risk of desiccation, it may, in fact, also be a cooling mechanism as 586 calories of heat energy are liberated in the phase change of state from liquid water in the plant to the gaseous form in the atmosphere. Alternatively, ice may act as an insulator when in contact with leaf tissue. A sheet of water on a leaf releases eighty calories of heat energy when the water freezes. That heat energy is conducted into the leaf and may sustain the tissue temperature at higher temperatures even as the ambient

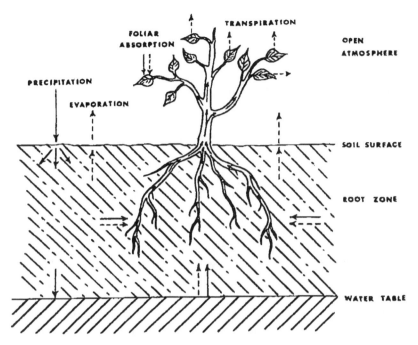

Figure 1. Soil-plant-atmosphere continuum (SPAC). As modified from Brown, R. W. 1977. Water Relations of Range Plants. In *Rangeland Plant Physiology* (Sosebee, R. E., Ed.). Society for Range Management. Denver, CO.

temperature drops below freezing. Combined with its high specific heat (the requirement of one calorie to raise one gram of pure water 1°C) these changes of state permit plants to regulate metabolism for an energetically preferred condition of homeostasis or constancy.

Water is often referred to as the universal solvent. Although this is not entirely correct, water does dissolve more substances than any other common liquid. Even so, it is still relatively inert and innocuous: it does not readily enter into chemical reactions. As such, it serves even more effectively as the matrix for cellular processes and, hence, the operation of whole organisms. This in turn regulates the patterns of plants and animals that characterize the Great Plains. For example, water regulates the distribution of tallgrass, mixed grass, and shortgrass prairie on the landscape.

5

AVAILABILITY OF WATER

Precipitation ranges from a high of 90 cm (35 in.) in the east to 50 cm (10 in.) or less in the west across the Great Plains. Also, temperature ranges broadly with the mean annual temperature near 15°C (60°F) in north Texas to about 7°C (45°F) in North Dakota. The consequence of this is that vegetation and land use patterns vary across the Great Plains. The potential natural vegetation includes 1) the tallgrass prairie, 2) northern mixed grass prairie, 3) southern mixed grass prairie, and 4) the shortgrass steppe (French, 1979). Water is the key regulatory resource or controller of vegetation type across the Great Plains. Vegetation in turn forms the template for the sorts of animals that inhabit a region and the associated activities of the humans who call this region home.

As precipitation decreases from east to west, it also becomes more variable. It is common for regions of the shortgrass steppe through eastern Wyoming, Colorado, New Mexico and western Nebraska, Kansas, Oklahoma, and Texas to receive nearly 80 percent of the annual precipitation in brief intense events, as in cloudbursts; in just a few short weeks in the spring; and again in the fall. Grasses are largely short and drought tolerant and include *Buchloe dactyloides* (buffalograss) and *Bouteloua gracilis* (blue grama). The low and erratic pattern of precipitation also restricts agriculture in the western region of the Great Plains to cattle and sheep grazing, with some production of drought-tolerant row crops such as wheat, sunflowers, and sorghum. Alternatively, the central portion of the Great Plains— through central North and South Dakota, central Nebraska, Kansas, Oklahoma, and north-central Texas—receives enough precipitation for production of row crops such as cotton in the south and corn and soybeans in the north. Mid-height grasses characterize the mixed-grass prairie including the cool season *Agropyron smithii* (western wheatgrass) and *Koeleria pyramidata* (prairie junegrass) in the northern mixed-grass prairies of eastern Montana, Wyoming, and western North and South Dakota. In the southern mixed-grass prairies of central Kansas, Oklahoma, and Texas, warm season mid-height grasses predominate including *Andropogon scoparius* (little bluestem) and *Bouteloua curtipendula* (sideoats grama). To be sure, rivers that flow across the Great Plains from the Rocky Mountains to the east and the underlying Ogallala Aquifer afford farmers the opportunity to nourish their crops with surface water and groundwater irrigation. Finally, in the eastern portions of the Great Plains, rainfall may be adequate for dryland production of crops such as corn and soybeans.

Although the global water budget is balanced—it cannot be otherwise because our earth is a closed system—there is no question that there are regional imbalances. Ours is an earth system that relies on radiant energy from the sun for its continual function. All cycling of essential elements is earthbound with no significant extraterrestrial exchanges. Routinely, output of water exceeds input across the Great Plains (French, 1979). Temperature, however, can dramatically influence the availability of water for plant growth and development. For example, precipitation in central North Dakota may be 65 cm (25 in.) per year with a mean temperature of about 5°C (40°F), while central Oklahoma may receive 90 cm (35 in.) of precipitation per year with a mean temperature of about 15°C (60°F). Yet, the dominant potential natural vegetation from north to south in the tallgrass region of the Great Plains is very similar. This 1000-mile north-south transect is characterized by a composite of warm and cool season grass species. Among the warm season grasses are *Andropogon gerardii* (big bluestem), *Sorghastrum nutans* (Indiangrass), *Spartina pectinata* (prairie cordgrass) and *Sporobolus asper* (tall dropseed). Cool season grasses include *Elymus canadensis* (Canada wild-rye), *Poa pratensis* (Kentucky bluegrass), *Phleum pratense* (timothy), and *Dactylis glomerata* (orchardgrass). How is it that the herbaceous natural vegetation can be so similar in North Dakota and Oklahoma? In fact, although Oklahoma receives 40 percent more precipitation than North Dakota, it is also 50 percent warmer on average. Consequently, although *total* precipitation is higher in Oklahoma, *effective* precipitation for plants is nearly the same as in North Dakota. The combination of evaporation and transpiration from plants in Oklahoma causes only about 75 percent of the 90 cm of precipitation to be available for plants—about the same as the total precipitation in Oklahoma.

Other than precipitation, the two key water resources for human activity are surface water and groundwater. Several significant rivers dissect the Great Plains. The Missouri River flows largely from northwest to southeast across the northern tier of the Great Plains.

7

The North, South, and Platte Rivers (in Colorado, Wyoming, and Nebraska) along with the Arkansas River (in Colorado and Kansas) flow from west to east across the central region of the Great Plains. The Canadian and Red Rivers also flow primarily from west to east through Oklahoma and Texas.

CENTRAL NEBRASKA AS A CASE STUDY

Surface Water

The Platte River is a representative model for issues of multiple-use pressure on water resources. The Big Bend Region of the Platte River through central Nebraska, an approximate 120-mile stretch of river, experiences intense multiple-use pressures. This region is a focal point for urban, agricultural, industrial, and wildlife stakeholders in relation to water quality and quantity. Nebraska law informs priority use and associated decision-making. Municipal use receives first priority followed by agricultural and industrial use. The Endangered Species Act (ESA) has thrust the federal government into the decision-making arena because there are at least three bird species on the endangered list in the region: the least tern, the piping plover, and the whooping crane. Beginning in July 1997, the states of Nebraska, Colorado, and Wyoming have been charged with negotiating a viable plan for water allocation for the endangered species along with the long-term legal status in Nebraska for municipal, agricultural, and industrial use. Under the auspices of a three-state Cooperative Agreement, the negotiation of such a plan has proved to be a challenge. The historical model of "first in time, first in right" for water extraction from the Platte River system for "beneficial use" has been decidedly challenged. Further, as the three-state effort takes shape, it has become politically clear that the hydrological reality of the connection between surface water and groundwater must be recognized. The state of Nebraska has made such a legal provision in recent years and is presently assessing mechanisms for quantifying that connection.

Groundwater

There are about 18,000 groundwater wells across the Big Bend Region of the Platte River in Nebraska (Central Platte Natural Resources District, 1999). Most of the row crop production, corn and to a lesser extent soybeans, depends on groundwater for irrigation. There is some surface river water from canals and reservoirs that supplements groundwater for irrigation purposes.

The groundwater supply for agricultural production is routinely considered to be very good. The Ogallala Aquifer underlies this region extending under fully two-thirds of Nebraska. The aquifer extends just into South Dakota to the north and as far south as the Oklahoma Panhandle and west Texas. Although portions of the aquifer have irreversibly dried up in the south and west, the depth to groundwater remains only about four to thirty-five feet in the Platte River valley. In fact, there are areas in the rainwater basin to the south of the Platte River that have shown a consistent increase in the groundwater supply. This is very likely a consequence of the surface water–groundwater linkage and the return rate from the river to the aquifer supply because more water has been made available in the river in recent years for wildlife habitat. This has partly been a consequence of the cooperative agreement among Nebraska, Colorado, Wyoming, and the Department of Interior, which is designed to allocate in-stream river flows for endangered species.

Multiple-Use Pressures

Demand for water in the Big Bend Region of the Platte River reflects conditions throughout the Great Plains. Although some rural areas have experienced a decline in population in recent decades, the urban areas continue to expand, and the municipal and industrial demand has correspondingly increased. Further, water demand for irrigation will continue to increase as long as the profit margin is favorable for key commodities like corn. Hence, there will continue to be intense multiple-use pressures on the key resource of the Great Plains—water.

A WATER ETHIC

Ordinarily, discussions about environmental ethics tend to focus on landscapes that are perceived to be spectacular or at least rare in nature: national parks, mountain ranges, seashores, and wild rivers. Although the Great Plains may seem at first glance to harbor little that is rare, there are many facets of the region that are at risk. This includes both natural and human systems—ecosystems and agroecosystems with the associated human community structure. Water availability forms the underpinning of decision-making about ecosystems and agroecosystems throughout the Great Plains. Although some people have contended that there are few organisms that are rare, let alone endangered across the Great Plains, it is clear that our perceptions of sustainability change as more is learned about the region. Further, it also makes eminent sense to protect and conserve ecosystems and agroecosystems while the elements therein remain relatively common, and not to wait until they are on the brink of extinction.

This sense of commonness should not be interpreted to mean commonplace. "Commonplace" is defined as ordinary, undistinguished, or uninteresting. The ecosystems of the Great Plains are none of these. Recognizing the worth of the ecosystems of the Great Plains requires establishing a mature environmental ethic. The ecosystems of the Great Plains—grasslands, including the tall-, mixed-, and short-grass prairies; woodland savannas; riparian corridors; and the myriad microhabitats—all have distinct quantitative features that make them special. These include the dynamics of energy transfer, nutrient cycling, and species' diversity and density. Unlike the visceral and emotionally based stirrings that may accompany the view of a snow-crested mountain, the undulating prairie grasses, or a sunset on a secluded beach, insight is acquired, not innate. That is, the human sense of the worth of nature requires that we exercise the cognitive portions of our minds in sifting all the bits of information, and misinformation, in the world around us. We must become students of nature. Most models of the valuation of nature encompass utilitarian, aesthetic, and intrinsic elements (Hargrove, 1989). Although the aesthetic and intrinsic value systems date to the seventeenth century, the utilitarian value system is largely a twentieth-century phenomenon. Economic valuation tends often to override aesthetic and intrinsic valuation in the United States. Because nature most certainly provides food, shelter, clothing, and medicine, Ehrenfeld (1976) says there is much more to the worth inherent in nature. In fact, if necessary, an economic valuation can be ascribed to "nonresources" such as those Ehrenfeld lists:

1. Recreational and aesthetic values
2. Undiscovered or undeveloped values
3. Ecosystem stabilization values
4. Value as examples of survival
5. Environmental and baseline monitoring values
6. Scientific research values
7. Teaching values
8. Habitat reconstruction values
9. Conservation value (avoidance of irreversible change)

This sort of list as a protocol for economic decision-making most assuredly would place the private individual or corporation at risk, with bankruptcy as a likely outcome. However, Dubos (1976) promotes a kind of stewardship that is driven by a mutualism between humans and nature that neither overemphasizes preservation of earth's resources like museum pieces nor advocates indiscriminate natural resource exploitation. An important link must be forged not to economic valuation of natural resources, which we have certainly mastered, but rather to the human psyche and our place in nature.

Scientific inquiry is a search for truth. A part of this quest is the power that accompanies knowledge: the knowledge we have to reflect on our role in the global system (Bicak, 1997). Regardless of whether our personal models are economically, aesthetically, or intrinsically based, there ought to emerge in the natural worldview a sense of the ethical treatment of nature and natural resources. Development, then, of a mature environmental ethic causes key questions to come to mind.

- Is the price of gasoline really in keeping with its worth as a finite energy source derived from fossil fuels?
- Can we ascribe economic worth to regions of the earth about which we know very little, as in the invertebrates of tropical regions?

- How much is water really worth in the urban centers of the Great Plains, cities like Amarillo, Wichita, and Cheyenne?

Odum (1971) questions the wisdom in separating natural currencies of ecosystems from economic currencies. For example, the calories or joules that characterize energy flow in the grasslands of the Great Plains defines them in terms of key properties like stability and sustainability. Whenever the economic currency, like dollars, diverges greatly from the ecological currency, like calories, the sustainability of the natural system is tenuous. This sets the stage for an ethical attitude of competition rather than cooperation between human aspirations and nature. It is often said that necessity is the mother of invention. It has become increasingly clear in recent years that the science of water must dovetail with economic, political, and cultural belief systems in order for a sound environmental ethic to emerge in relation to issues of water quantity and quality in the Great Plains.

A water ethic is emerging across the Great Plains that recognizes the importance of balancing three types of sustainability: ecological, agricultural, and cultural (Douglass, 1986).

Ecological Sustainability

Ecological sustainability requires an ethic that not only recognizes but abides by the fundamentally important notion that there are physical limits on natural resources such as water that ought to inform decision-making (Jackson and Piper, 1989). For example, simply because an extension in agricultural or urban land use is technologically and economically feasible in the short term, it is not sufficient rationale for disturbance or destruction of a natural system. Long-term consequences may be economically as well as ecologically disastrous. A case in point is the legacy of failed center-pivot irrigation systems across the Great Plains that were installed on highly erodible lands in the early 1980s. Although there was great economic promise at the time with high commodity prices, this soon vanished as the soil and water resources in fringe arable areas were depleted.

Agricultural Sustainability

Agricultural sustainability may be viewed from a strictly economic perspective: if supply matches demand at a "reasonable" price to ensure "normal" profits, then the agricultural system is sustainable. Others recognize, however, that agricultural sustainability requires an additional dimension—conservation of finite resources like soil and water. To that end, agricultural conservation practices have decidedly changed for the better in many regions of the Great Plains. Among these practices is the appropriate use of pivot irrigation systems to enhance water use efficiency and lessen groundwater contamination with nitrate. In many areas, gravity irrigation has been replaced by the use of gated pipe, which also improves water use efficiency.

Contour farming, seasonal alternation of crops, minimum tillage practices, and other cultivation practices that reduce soil compaction all combine to increase the odds for long-term sustainability of agriculture across the Great Plains.

Cultural Sustainability

Cultural sustainability is not easily defined. For many people across the Great Plains the definition of community or cultural sustainability centers on the quality of life—real or perceived—associated with the "family farm." Coupled with the family farm is the network of towns and counties that link the agricultural sector to the urban environment. There are presumed values associated with family and kinship that underscore a need to preserve the rural way of life as it dates in some ways to the Homestead Act in 1862.

Economic hardships, certainly related at least in part to water availability, have led to declining population in many areas of the Great Plains. Indeed, it has been proposed that vast expanses of the Great Plains, particularly in the northern tier, be returned to an unfarmed and unpopulated condition: the Buffalo Commons plan (Popper and Popper, 1987).

The economic valuation of water and its true worth as a finite natural resource has long been divergent (Odum, 1971). That is, for many decades water for all users—municipal, agricultural, industrial, and wildlife—has either been undervalued or subsidized. On occasion, the subsidy may seem rather indirect, as in export-import regulations and elements of the prevailing farm legislation, but very often the direction of activity on the Great Plains has been tied to the reality of a limited supply of water. In recent years there appears to be developing some convergence of the economic valuation and the true worth of water. This may be driven by economic and ethical considerations and would appear to be an indicator of some optimism in the decades ahead for all three types of sustainability: ecological, agricultural, and cultural.

THE FUTURE

Rivers and streams typically make up less than 5 percent of the landscape across the Great Plains. Water in the Ogallala Aquifer is an integral factor in irrigated row crop production across the Great Plains. Precipitation varies across the region. Overall, water availability constrains human development and is the defining controller of ecosystem structure and function.

Multiple-use pressures, including municipal, agricultural, industrial, and wildlife demands, have only increased in recent years. The number and variety of stakeholders, those with a vested interest in Great Plains water issues, continue to increase. Efforts such as the three-state Cooperative Agreement among Nebraska, Colorado, and Wyoming are designed to reconcile the disparate priorities regarding water allocation (Cooperative Agreement, 1997). Ranchers, farmers, urban dwellers, state and federal agency representatives, and members of environmental organizations such as The Nature Conservancy and the National Audubon Society, are all intensely concerned with the fate of that simple, yet remarkable and absolutely essential substance—water.

Although it is unclear how the future of the Great Plains will unfold, there is no question that water issues will be at the center of decision-making. An ongoing sense of awareness and subsequent urgency regarding Great Plains water issues has flavored developments on the Great Plains for centuries. That is not about to change any time soon.

REFERENCES

Bicak, C. J. 1997. The Application of Ecological Principles in Establishing an Environmental Ethic. *The American Biology Teacher* 59, no. 4: 200–206.

Central Platte Natural Resources District. 1999. *Comprehensive Resources Plan.* Developed in accordance with Nebraska Law (Section 2-3276). Ron Bishop, Manager. Grand Island, Nebraska.

Cooperative Agreement. 1997. *Basinwide Recovery Program for Endangered Species in the Central Platte River Basin.* Implementation among the states of Nebraska, Wyoming, and Colorado and the U. S. Department of the Interior. Dale Strickland, Executive Director.

Douglass, G. K. 1986. "Sustainability of What? For Whom?" Paper presented at the Symposium on Sustainability of California Agriculture. Sacramento, California.

Dubos, R. 1976. Symbiosis Between the Earth and Humankind. *Science* 193: 459–62.

Ehrenfeld, D. W. 1976. The Conservation of Non-Resources. *American Scientist* 64: 648–56.

French, N. R. 1979. *Perspectives in Grassland Ecology.* New York: Springer-Verlag.

Hargrove, E. C. 1989. *Foundations of Environmental Ethics.* Englewood Cliffs, N.J.: Prentice-Hall.

Jackson, W., and J. Piper. 1989. The Necessary Marriage Between Ecology and Agriculture. *Ecology* 70, no. 6: 1591–93.

Odum, H. T. 1971. *Environment, Power, and Society.* New York: Wiley-Interscience.

Popper, D. E., and F. J. Popper. 1987. The Great Plains: From Dust to Dust, a Daring Proposal for Dealing with an Inevitable Disaster. *Planning* 53: 12–18.

Hydrological Drought as a Settlement Inhibiting Factor

Conrad T. Moore

For more than two-thirds of a century historians, geographers, and other scholars have debated the issue of why the Great Plains was passed over as a region for settlement until relatively late in the nineteenth century. Prior to the 1870s, hundreds of thousands of immigrants traveled across the plains en route to Oregon, California, Utah, and other destinations in the Far West. Few had any inclination to cut short their journeys and establish homesteads in the Great Plains.

At the forefront of these discussions has been the importance of the image of the region, or extensive tracts within it, as a "desert." This assessment was first introduced by members of the Lewis and Clark expedition during the hydrological drought that prevailed on their journey up the Missouri River and across the northern plains in 1805. The following year, Zebulon Pike provided a similar description of the central plains during his ascent of the Arkansas Valley, although drought was not a factor in this case. In 1820, members of the Stephen H. Long expedition traveled up the Platte and South Platte Valleys to the Rocky Mountains and returned via the Arkansas and Cimarron Valleys. Despite the frequent and copious rains that kept the Long expedition from fording the North Platte River on their way to the South Platte, and bogged down Captain John Bell's horses in the muddy floodplain of the Arkansas River, Major Long labeled the region the "Great Desert" on his map.

Edwin James, the chronicler of the Long expedition, also described the region as a "desert" and extended it into the northern plains on the basis of his familiarity with the journals of the Lewis and Clark expedition. From William Clark's and Sergeant John Ordway's descriptions of the northern plains as "Deserts of America" and the "Deserts of North America," respectively, in 1805, and Major

Long's description of the "Great Desert" emerged the amalgamated concept of the "Great American Desert" that for years thereafter was adopted and perpetuated in school textbooks and in the minds of many who subsequently crossed the plains. Ultimately, Major Long's "Desert" was displaced by "Plains" with "Great" persisting to the present. The tremendous variation in the actual climatic conditions that prevailed during the formative period of the "Great American Desert" concept, ranging from extremely dry to the inordinately wet, has generated a considerable body of contemporary literature dealing with existing conditions and regional assessments.

In the period from the early 1930s through the mid-1970s, most scholars avoided the issue of climatic conditions, choosing instead to attribute the "desert" image of the plains to the lack of trees. This was an unquestionable environmental deficiency, much more so during the first three-quarters of the nineteenth century than at present, because of extensive depletion of the riparian cottonwood forests by the Plains Indians for winter feed for their horse herds and for fuel (Moore, 1999). As Webb (1931) pointed out, "Had the region been heavily forested and well watered, there never would have been an Oregon Trail, for there would have been no reason for one." Smith (1950, 149) noted that the absence of trees held much more significance than merely the lack of construction lumber and fuel. "Americans were used to judging the fertility of new land by the kind of trees growing on it. . . . The absence of trees over great expanses of the plains was regarded as proof that the area was unsuited to any kind of agriculture and therefore uninhabitable by Anglo-Americans" (202–3). Essentially the same view was expressed by Hollon (1975): "The absence of trees in large areas of the Great Plains indicated to each traveler that the soil was sterile and incapable of growing crops even with sufficient water" (66).

In the 1970s, the focus of regional explanation for the desert image began to shift toward the importance of climatic conditions. Despite his previous disavowal of the importance of water, Hollon also stated in 1975 that "[a] humid cycle followed the Civil War, and the pioneers then moving westward altered their concept of the desert. . . . By 1870 the concept that the Trans-Mississippi West constituted a vast and trackless desert had disappeared almost entirely from official maps and school geographies" (11–2). Saarinen (1977) emphasized the importance of "the broad seasonal and cyclical fluctuations in weather and climate . . . where observers in different places and at different times would have been reporting on entirely different conditions, though all within the Great Plains" (296).

In an effort to resolve the dilemma posed by Saarinen, dendroclimatologists have determined from tree-ring chronologies on the periphery of the plains that drought during the period 1855–1865 was at least as severe as that of the 1930s (Meko, 1992; Stahle and Cleaveland, 1988; Stockton and Meko, 1983).

Despite Hudson's (1996) conclusion that nineteenth-century overland travelers "were reasonably faithful in describing what they must have seen" (3) both with respect to drought and periods of unusually heavy precipitation, historical geographers have, in contrast, found little evidence of drought and the problems it would have posed for those crossing the plains from the 1820s through the 1860s. This led Allen (1985) to conclude that the primary reason that immigrants did not settle on the plains during the peak migration years of 1845–1860 was not that they viewed the region unfavorably, but that "what lay beyond was even better" (212).

Mock (1991) suggested a partial explanation for this apparent contradiction when he noted that "many parts of the Great Plains are devoid of trees more than a hundred years old, making dendroclimatic applications impossible" (28). He also pointed out from his own comparative analysis "that dendroclimatic and instrument data do not always agree on drought occurrence" (44). However, with regard to future research he concluded that "extraction of climatic information from all proxy and historical sources is far from being accomplished, and eventually will give researchers a full understanding of drought and climate in the Great Plains prior to the twentieth century" (52). The purpose of this chapter is to contribute toward that end and determine from these findings the extent to which drought affected traveler assessments of the settlement potential of the region.

METHOD OF ANALYSIS

If overland travelers were reasonably accurate in describing what they saw and if they encountered severe or extreme drought conditions in the mid- to late 1850s and 1860s or at other times, it is unlikely that most would have described the plains in favorable terms with respect to settlement potential. Regional emigration during the 1930s clearly suggested the opposite.

Part of the problem with respect to previously published historical geographic studies is that they have been based on relatively limited numbers of travel accounts. Only one examined more than a hundred accounts, and that pertained exclusively to Mormon immigrants who were so caught up in a state of religious euphoria that they described grassland devastation by locusts in positive terms (Jackson, 1975).

For this chapter, 364 overland travel accounts written during the years 1805–1880 were examined for evidence of drought; its impact on travelers and their draft, pack, and riding animals; and assessments of the desirability of the region for settlement. Overland travel was very limited prior to 1819 and declined rapidly after 1869 with the advent of transcontinental rail travel. With the exception of the 1805–1818 period, observations were recorded for all years, although a few involved very restricted areas (e.g., the environs surrounding a fur-trading post). The other geographic limitation was that there were considerably more observations in the central plains than in the northern and southern regions simply because the Platte, Arkansas, and other west-to-east flowing rivers provided the most direct and easy access to western destinations.

In order to eliminate possible misinterpretation by observers concerning the causes of grass scarcity, only droughts with distinct hydrological effects were considered. The criteria established by the National Climatic Data Center of groundwater supply and streamflow were used in this analysis. Overland travelers reported dry springs, dry perennial tributaries, very low water levels in rivers, and dry riverbeds, conditions indicative of severe or extreme hydrological drought during twenty of the seventy-one years in which observations were reported. With the exceptions of 1822, 1830, and 1874, droughts were documented by at least two observers and were never contradicted by other observations.

HYDROLOGICAL DROUGHTS AND THEIR IMPACTS

Documentary Evidence

Half of the hydrological drought years were observed during the forty-six-year period 1805–1850, with four of these occurring during 1841–1850. Seven hydrological drought years were documented during the 1855–1865 period, thus substantiating the conclusion arrived at by dendroclimatologists. The problem with respect to the dendroclimatic studies is that only three of the remaining thirteen hydrological drought years that were substantiated by first-hand field observation (i.e., 1822, 1842, and 1874) matched the findings from tree-ring analyses.

There are several possible explanations for these differences. Droughts that were observed in 1833 and 1877 may not have been detected because they were too remote from sites from which modern tree-ring chronologies were constructed. The drought in 1830 in eastern and central Kansas was observed in the fall and may have begun after the spring and early summer period of primary tree-ring growth. The droughts of 1845, 1846, 1850, and 1853 all shared a common precondition. Very wet years were observed in 1843, 1844, 1849, 1851, and 1852. This suggests the possibility that surplus soil water and groundwater accumulations during the wet years may have masked the effects of drought for the deep-rooted post oaks and ponderosa pines used in the dendroclimatic reconstructions. In the past century, 1909, 1926, 1951, and 1974 stand out as years of severe or extreme hydrological drought in which preceding wet conditions apparently obscured dendroclimatic results (Karl and Knight, 1985). Reciprocally, for the droughts observed in 1822, 1842, 1874, and during the period 1855–1865, only 1863 was preceded by a comparably wet year. In that instance, however, the extreme drought years of 1859 and 1860 followed by relatively dry conditions in 1861 would have created such a soil moisture and groundwater deficit that there would have been little chance of very wet conditions in 1862 resulting in a drought-masking moisture surplus.

Although field observations were lacking for the northern plains prior to the fall of 1804 and for the southern plains in 1818 and 1833, consensus dendroclimatic evidence from ponderosa pine and post oak chronologies indicates that 1804 and 1833 were also very wet years (Stahle and Cleaveland, 1988; Stockton and Meko, 1983). If this is so, it would explain why the extreme hydrological droughts that prevailed across the northern plains in 1805 and in Oklahoma in 1834 went undetected despite the ponderosa pine and post oak chronologies from these regions. Lack of detection of the extreme hydrological drought of 1819, when water levels in the Arkansas River were at least as low as in 1860, defies explanation. Dendroclimatic evidence indicates that moisture conditions for each of the two preceding years were average or only moderately above average.

For the ten hydrological droughts that were observed during the first half of the nineteenth-century, water deficits, for the most part, were not as pronounced as in 1857–1874. In only one instance (1805) was a river channel reported as being dry or with water only in standing pools, whereas this situation was repeatedly documented for six of the seven drought years of the latter period. Very low water levels, however, were observed in all years, dry perennial tributaries for all years except 1830, and dry springs in 1819, 1842, and 1846.

Animal and Human Impacts

With the notable exceptions of 1834, 1845, and 1846, impacts on overland travelers and their transportation animals were not as devastating as in 1857–1877, when draft, pack, and riding animals perished or were humanely destroyed during seven of the eight drought years. In 1822, the fur trader Jacob Fowler resorted to digging holes in the sand of a dry perennial tributary to avoid perishing from thirst during a crossing from the upper Canadian River to the Arkansas River (Coues, 1898). In 1830, Isaac McCoy's surveying expedition in eastern and central Kansas had to return prematurely to Fort Leavenworth due to a lack of forage and water (Barnes, 1936). Similarly, the drought-imposed scarcity of water and grass in 1850 forced thousands of immigrants to turn back before they reached the forks of the Platte River (Williams, 1969).

During each of the three years previously mentioned, transportation animals died and humans experienced considerable suffering. In 1834, the horses of a military expedition returning across central Oklahoma perished for lack of water (Catlin, 1926). On this same journey, soldiers were forced to drink the urine- and feces-contaminated water in buffalo wallows because of extreme thirst. In 1845 and 1846, all of the plains tributaries of the Platte and Arkansas Rivers went completely dry—the result of, at least in some areas, a total absence of rain, as well as hot southerly "sirocco-like" winds and, in 1846, temperatures that reached 120°F (Calvin, 1951; Cooke, 1857; Parkman, 1917; Robinson, 1932). William Clayton, on arriving at Fort Laramie in early June, 1847, reported from a conversation with the principal trade officer that "they have had no rain for two years until a few days ago" (Clayton, 1973, 210). Those who ventured away from the floodplains of the Platte and Arkansas had to dig holes in the sand of dry perennial tributaries to obtain water (Cooke, 1857; Parkman, 1917). Horses, mules, and oxen belonging to immigrants and the military perished in large numbers (Cooke, 1857; Hughes, 1848).

Because of the prevalence of the fur trade during the first half of the nineteenth century, drought impacts on river transportation were more frequent and pronounced than during the last thirty years, when the fur trade faded into near obscurity. Four of the five references to interruption or termination of water-borne transportation involved the shallow-draft bullboats used by fur traders to convey their furs to St. Louis. In the summer of 1833, Nathaniel Wyeth and his party constantly ran aground while descending the Missouri from Fort Union to the vicinity of Fort Pierre. In 1842, 1845, and 1846, traders from Fort Laramie encountered even greater difficulties during their descent of the Platte, although their bull-boats drew only six to eight inches of water. In 1842, the fur traders were forced to abandon their boats seventy miles west of Grand Island. In 1845, they failed to reach the forks of the Platte, and wagons were sent out from Fort Laramie "to transport the packs back to the fort" (Cooke, 1857, 312). In 1846, the furs were cached some sixty miles west of Fort Kearny with the fur traders proceeding eastward on foot to obtain supplies (Bryant, 1848; Cooke, 1857; Sage, 1958).

The fifth instance documenting disruption of water transportation by drought in the first half of the nineteenth century was reported by the naturalist Thomas Nuttall, during his explorations in eastern and central Oklahoma in the spring, summer, and fall of 1819. Near Fort Towson in southeastern Oklahoma, he noted on June 6th, "All the lesser brooks and neighboring springs were now already dried up" (Nuttall, 1905, 280) No measurable rain was encountered from that date until he returned to Fort Smith on November 4. Concerning the Arkansas River he wrote, "no boats drawing more than 10 or 12 inches of water could possibly navigate it from the Dardanelles to the Verdigris" (Nuttall, 1905, 280).

In comparison with the extreme hydrological droughts that prevailed from 1857 through 1877, those of 1853 and 1855 were no more than severe. Dry springs and perennial streams were reported across all of Oklahoma in 1853, and levels in the Canadian River were very low (Foreman, 1941). In late August of 1855, William Chandless noted concerning the Platte River at Fort Kearny that "a few shallow streams of water hardly make their way through sand and shingle [i.e., gravel]" (Chandless, 1857, 63). Neither travelers nor their transportation animals suffered undue hardships, including those who traveled considerable distances between river valleys.

As previously suggested, the period 1857–1865 appears to have been the most drought-afflicted of the nineteenth century. Extreme hydrological droughts occurred during six of the nine years. Both the Cimarron River near the 100th meridian in 1857 and the North Fork of the Red River in southwest Oklahoma in 1859 had water only in standing pools (Bandel, 1932; Estep, 1960). The maximum water depth in the Platte, ninety miles west of Fort Kearny in late April, 1859, was reported as being "only shoe deep" (Baker, 1861, 9). In 1860, the Smoky Hill River in central Kansas and the North Canadian and Cimarron Rivers in northwest Oklahoma were completely dry, as was the Verdigris in southeast Kansas (Anonymous, 1958; Stuart, 1959). Navigation on the Arkansas to Fort Smith and beyond was suspended from May through September 1860 (Anonymous, 1958). In 1863, all of the eastern tributaries of the Missouri from Sioux City, Iowa, to Fort Pierre, South Dakota, had water only in standing pools, and riverboats on the Missouri were aground in the vicinity of Fort Randall (Goodwin, 1970; Peirce, 1963). In southern Texas in the fall of the same year, the San Antonio and Nueces Rivers were described as being "mere trickling threads of water with here and there a small pool" (Butler, 1925, 481). In the summer of 1864, both the Sheyenne and James Rivers in eastern North Dakota were "only pools of poor stagnant water" (Mattison, 1969, 219). In eastern Montana in August 1865, the Powder River was as dry as the name suggests and the Tongue River contained only standing water (Cole, 1961; Walker, 1961).

Horses, mules, and oxen perished or were humanely destroyed during all drought years of the 1857–1865 period, and travelers dug holes in the sand of dry river and stream channels to obtain water in 1860 and 1864 (Myers, 1971; Stuart, 1959). In the summer of 1865, soldiers of the Powder River Campaign suffered such extreme thirst that they drank the water in a buffalo wallow "although it was impregnated with the excrement of these animals" (Cole, 1961, 70). In this instance, the expedition was forced to turn back before its mission was accomplished.

Although separated by sequential years of average or above average precipitation, the droughts of 1874 and 1877 were unquestionably at least as extreme as any of the years of the 1857–1865 period. In August and early September of 1874, Nelson Miles led a column from Fort Dodge, Kansas, to the Red River Valley in Texas while participating in the Red River Indian War. With respect to the southerly journey he wrote,

> In many places no water was to be discovered in the beds of streams, and only at long intervals were there found stagnant holes containing some, often impregnated with gypsum. Men rushed in frenzy and drank, only to find their thirst increased rather than slaked . . . The heat was almost unendurable, the thermometer ranging above 110 degrees in the shade, daily. (Miles, 1969, 166–167)

Subsequently, he added,

> On reaching the bed of the Red River, which at that point was nearly half a mile wide, there was only found a small pool of saturated gypsum and alkali, the stagnant water being rendered utterly unfit for use. During the chase the men tried every means of finding water, but without avail, and suffered so greatly that some of them resorted to the extreme of opening the veins of their arms and moistening their parched and swollen lips with their own blood. This expedient to relieve extreme suffering has occurred on two different occasions in my commands; at this time on the Red River of Texas, and again on the arid plains of Arizona. (Miles, 1969, 162)

As bad as the experience was for the soldiers participating in the Red River Indian War, the catastrophe that befell another military expedition from Fort Concho (San Angelo) in July and August,

1877, while pursuing Indians in west Texas was unsurpassed in the presettlement period. As a result of the extreme hydrological drought, four soldiers and an army scout died of thirst, as did twenty-three of the twenty-five horses. Probably all would have perished had the soldiers not resorted to drinking their own urine and ultimately the blood of the horses as they gave out, and had a rescue party from the fort not arrived in time (Cooper, 1940; King, 1940).

ASSESSMENTS OF SETTLEMENT POTENTIAL

Although emigrants from the humid eastern United States reported occasional drought periods when springs and perennial tributaries went dry and when water levels in rivers became very low, these did not occur with either the frequency or the severity of those they encountered in the plains, nor did they persist as long. Furthermore, none of the emigrants had ever seen a major river with a dry bed, nor had any of them experienced the consequences of the hydrological droughts previously described.

For those who recorded their assessments of the region (and many did not, including the two army officers who wrote of their travails on the southern plains in 1877), distinctions were usually drawn, either formally or tacitly, between the lands to the east of the 98th or 99th meridians and the more arid region lying to the west. Only on rare occasion were areas to the east described as a desert or in other unfavorable terms, irrespective of whether drought conditions prevailed or not. Despite the problems created by drought, most observers regarded the subhumid lands lying to the east as satisfactory for settlement. Isaac McCoy's 1830 assessment in northern Kansas near the 98th meridian was typical.

> Since we came into the vicinity of the Republican, or Pawnee River, wood has been more scarce than previously. The creeks, however, are all wooded. Fuel would be sufficient for a considerable population—chiefly elm, cottonwood, and willow near the rivers—farther from the rivers is more wood on the creeks, and of different kinds . . . The country *is* habitable thus far. (Barnes, 1936, 372)

G. K. Warren's conclusion twenty-seven years later, after experiencing the droughts of both 1855 and 1857, was similar. With respect to the Nebraska Territory (which at the time also included Montana, Wyoming, and most of the Dakotas) he stated that "there are fertile tracts as far west as the 99th meridian, in the neighborhood of streams that are valuable, and contain wood enough to support settlements" (Warren, 1859, 541).

Negative Environmental Assessments

Apart from the six individuals who used the term "desert" to describe areas that were deserted or uninhabited, and the twenty-four observers who referred to only relatively localized areas crossed during one or two days of travel as environmental deserts or as "barren," "desolate," or "sterile" tracts, 129 individuals recorded negative assessments of the Great Plains during the years 1805–1880. Beginning with the references to the northern plains desert recorded by members of the Lewis and Clark expedition (Thwaites, 1904), these assessments extended through 1876. There were major differences, however, in the incidence of negative assessments during hydrological drought episodes as opposed to nondrought years and, with respect to the latter, the years prior to 1851, in comparison with the last thirty. During the hydrological drought years, 59 percent of the accounts contained negative assessments of the plains and there was no significant difference between the first half of the nineteenth century and the last three decades of the nineteenth century in the incidence of either the use of the term "desert" or other negative assessments.

For the remaining fifty-six years during which conditions ranged from moderately dry to very wet, there were three and one-half times more negative assessments during the first half of the nineteenth century than over the last thirty years of the nineteenth century, when the incidence of negative assessment dropped to less than 10 percent.

The description by members of the Stephen H. Long expedition of the plains or extensive areas within it as a desert, despite the copious rains and flooded river valleys, was not unique. Eleven other travelers prior to 1851 described the region as a desert during the very wet years of 1839, 1843, and 1849, with four using the formal name "Great American Desert." Surprisingly, only five cited an absence of trees as a criterion for their desert designations. After 1850, three others continued this pattern, concluding with Lieutenant Charles

King who, during the very wet summer of 1876, likened the Powder River Valley to the Sahara.

Overall, and irrespective of whether conditions were very wet or moderately dry, the most frequently cited criteria for negative assessments of the plains during nondrought years were the absence of trees and the presence of sand hills and plains. The latter appears to have been, at least in part, a legacy of the environmental descriptions provided by Zebulon Pike in 1806 and Edwin James in 1820. Pike wrote of his westward journey up the Arkansas Valley:

> These vast plains of the western hemisphere, may become in time equally celebrated as the sandy deserts of Africa; for I saw in my route, in various places, tracts of many leagues, where the wind had thrown up the sand, in all the fanciful forms of the ocean's rolling wave, and on which not a speck of vegetable matter existed. (Jackson, 1966, 2:27)

James's delimitation of the "Great Desert" was similarly based. "The Rocky Mountains may be considered as forming the shore of the sea of sand, which is traversed by the Platte, and extends northward to the Missouri, above the great bend" (Thwaites, 1905, 248). It is unclear, however, why James made this assumption for the northern plains because none of the members of the Lewis and Clark expedition cited the presence of sand hills or plains as a basis for their desert designations.

Under moderately dry conditions when the incidence of negative assessment increased to 44 percent, a third element was added—the sparse growth of desiccated grass. John K. Townsend's description of the country west of the forks of the Platte in May 1834 was typical.

> Instead of the extensive and apparently interminable green plains, the monotony of which had become so wearisome to the eye, here was a great sandy waste, without a singe green thing to vary and enliven the dreary scene . . . the straggling blades of grass which found their way to the surface were brown and withered. (Townsend, 1905, 173–175)

During the hydrological drought years, the lack of grass, water, and precipitation accounted for 72 percent of the environmental deficiencies cited by those recording desert and other negative regional descriptions. An absence of buffalo and other ungulates was also mentioned on numerous occasions. Coupled with these deficiencies during four of the years were very high temperatures, hot dry winds, sand storms, and, in one instance on the southern plains, dust storms that matched the worst of the 1950s (Moore, personal communication). Two of the enlisted men documented the suffering experienced by the Army of the West during its march up the Arkansas Valley and across southeastern Colorado in the summer of 1846. Near Bent's Fort, Sergeant Jacob Robinson wrote:

> The country becomes a desert, extremely hot; the wind blows from it as from a heated oven, causing soreness of the eyes and bleeding at the nose. On the 4th of August we left the Arkansas . . . the heat intense, thermometer 120; clouds of dust almost suffocate the men, who are in confusion and grumbling. Water-mirage appears, but no water; and last and worst the dreaded Sirocco or hot wind blows, which burns us even through our clothes. (Robinson, 1932, 19–20)

With reference to this same tract, Private John Hughes added:

> We suffered much with heat, and thirst, and the driven sand—which filled our eyes, and nostrils, and mouths, almost to suffocation. Many of our animals perished on the desert . . . The Roman army under Metellus, on its march through the deserts of Africa, never encountered more serious opposition from the elements that did our army in this passage over this American Sahara. (Hughes, 1848, 62–63)

Temperatures were also very high during the summers of 1859 and 1860. In late July, 1859, Lieutenant William Burnet wrote from his post in north-central Texas, "I never have seen such hot weather, as we have had here this summer: The thermometer has averaged 106 degrees in the shade from 10 o'clock in the morning until 5 o'clock in the evening, for the last month" (Estep, 1960, 376). Three months prior to this he had documented the most severe dust storm reported for the first three-fourths of the nineteenth century.

> Yesterday there came up a most singular storm, a very strong wind and a tremendous cloud of dust: about four o'clock in the afternoon it became dark enough to have lights, the sun was entirely obscured: the dust must come from [the] Llano Estacado. (Estep, 1960, 289)

As a result of these experiences and the fact that many of the army oxen died en route to Fort Cobb in western Oklahoma in August, Lieutenant Burnet provided the following discourse.

Our people won't learn by example or experience. The long wars the French carried on in Algeria cost vast amounts of blood and treasure; and, on a larger scale, they are the same as our Indian Wars on the Plains: The same half desert and entire desert country and the same roving, crafty enemy capable of subsisting on almost nothing and of moving with a rapidity almost incredible and continuing their marches to almost any distance . . . I think it would be as well to let the Indians have it for white men can never make a living here unless at the cost of great labor and money . . . There is but little water and that is generally unfit for use. (Estep, 1960, 380)

As bad as these conditions were, they only worsened in 1860 with, on a widespread regional basis, what appears to have been the most devastating drought of the nineteenth century. In July, Major John Sedgwick wrote,

We have marched five hundred miles since leaving Pawnee Fork, passing through the heart of the Kiowa range, and over the worst country I have ever seen. There has been no rain on it since last fall and consequently all the small streams and pools where water is generally found, were dry. (Hafen and Hafen, 1959, 203)

On November 17 at Fort Wise on the Arkansas River, just west of Bent's Fort, he added: "Today there is a drizzling rain, the first that we have had since reaching the post, over two months since" (Hafen and Hafen, 1959, 276). Temperatures during the summer were even hotter than in 1859. At Fort Cobb, Lieutenant Burnet wrote: "It has been very hot here, 115 degrees has been the highest the thermometer has reached but 110 degrees has been quite common" (Estep, 1961, 29). During a brief visit in eastern Kansas in late 1860, Charles M. Clark noted that "[a] prominent citizen of Seneca, a lawyer . . . stated that the heat during the summer months had been excessive, the thermometer often marking 119 degrees in the shade" (Clark, 1861, 9). The "sirocco-like" winds also returned. During his journey along the North Fork of the Canadian River, an anonymous enlisted man in Captain Sturgis's command wrote: "A hot, scorching wind blew across the prairie all day long, nearly burning all the skin off our face and hands" (Anonymous, 1958, 405). Later, he added: "The heavy drought that prevailed in Kansas the past summer, has caused a great many to abandon their homes on the frontier for homes farther east, where they could gain a livelihood during the coming winter" (Anonymous, 1958, 413).

Positive Environmental Assessments

Although a few desert assessments persisted during unusually wet years after 1850, preponderant views began to shift toward positive assessments of settlement potential, particularly with respect to the potential for livestock grazing. In 1851, the Jesuit missionary Jean Pierre DeSmet was very favorably impressed with the abundance of grass and water in the valleys of the Tongue and Powder Rivers that had resulted from "a wet spring" (DeSmet, 1972). In the southern plains, heavy precipitation in 1854 across both the Llano Estacado and Chihuahuan Desert areas of west Texas inspired Andrew Gray to write: "There are vast fields of fine grazing lands . . . covered with mezquit [sic] and grama grasses, of great exuberance, which retain their nutritious qualities through winter and summer" (Gray, 1963, 87).

As Hollon suggested, a predominantly wet period followed the Civil War. Travelers encountered frequent and heavy rains in 1866, 1867, 1869, and 1870 and, as a result, positive assessments prevailed. In 1866, James Rusling predicted that the western portion of the central plains "will yet become the great stock-raising and dairy region of the country" (Rusling, 1877, 232). In June 1869, Harriet Bunyard mentioned the "Pecos Desert" before crossing the region, but subsequently rejected the notion that the region was a desert due to unusually heavy precipitation and abundant grass all the way to El Paso (Bunyard, 1980).

CONCLUSION

The evidence provided by those who traveled within or across the plains and experienced the consequences of water scarcity or absence during hydrological drought years strongly suggests that the conditions that prevailed were the dominant factor inhibiting settlement prior to the end of the Civil War and, on occasion, thereafter. Wood and most other necessities could be imported. Water could not.

Furthermore, arriving at one's destination was not a given. As a journalist for the *Rocky Mountain News* wrote in April 1860 concerning the much shorter, but largely nonriparian Smoky Hill Trail, only "the foolhardy and insane" would select that route to the Colorado gold fields (Hafen, 1942, 265). During nine of the twenty hydrological drought years, journeys were aborted due to lack of water. For the peak migration years of 1845 through 1865, there was more than a fifty-fifty chance that venturing into the plains would be attended by calamitous consequences. Hydrological droughts prevailed during eleven of the twenty-one years, and only in 1853 and 1855 were extreme hardships not recorded by overland travelers. However, widespread crop failure was reported for settlers in Texas in 1853 and among the Indians of the upper Missouri in 1855 (Marcy, 1866; Warren, 1856).

REFERENCES

Allen, J. L. 1985. The Garden-Desert Continuum: Competing Views of the Great Plains in the Nineteenth Century. *Great Plains Quarterly* 5, no. 4: 207–20.

Anonymous. 1958. With the First U.S. Cavalry in Indian Country, 1859–61. In *Letters to the Daily Times, Leavenworth*. L. Barry (ed.). *The Kansas Historical Quarterly* 24, no. 3: 257–84; no. 4: 399–425.

Baker, H. H. 1861. *Overland Journey to Carson Valley, Utah*. Seneca Falls, NY: F. M. Baker.

Bandel, E. 1932. Frontier Life in the Army, 1854–1861. In *The Southwest Historical Series*. Vol. 2. R. P. Bieber (ed.). Glendale, CA: The Arthur H. Clark Company.

Barnes, L., ed. 1936. Journal of Isaac McCoy for the Exploring Expedition of 1830. *Kansas Historical Quarterly* 5, no. 4: 339–77.

Bryant, E. 1848. *What I Saw in California Being the Journal of a Tour Across the Continent and Through California in the Years 1846, 1847*. New York: D. Appleton and Company.

Bunyard, H. 1980. Diary of a Young Girl. In *Ho for California! Women's Overland Diaries from the Huntington Library*. S. Myres (ed.), 198–252. San Marino, CA: Huntington Library.

Butler, P. B. 1925. Sixty-eight Years in Texas. In *Trail Drivers of Texas*. Vol. 1. J. M. Hunter (ed.), 479–84. Nashville, TN: Cokesbury Press.

Calvin, R., ed. 1951. *Lieutenant Emory Reports: A Reprint of Lieutenant W. H. Emory's Notes of a Military Reconnaissance*. Albuquerque: The University of New Mexico Press.

Catlin, G. 1926. *North American Indians*. Edinburgh, Scotland: John Grant.

Chandless, W. 1857. *A Visit to Salt Lake: Being a Journey Across the Plains, and a Residence in the Mormon Settlements and Utah*. London, England: Smith, Elder and Company.

Clark, C. M. 1861. *A Trip to Pike's Peak and Notes by the Way*. Chicago: S. P. Rounds.

Clayton, W. 1973. *William Clayton's Journal*. New York: Arno Press.

Cole, N. 1961. Report of Expedition. In *Powder River Campaigns and Sawyers Expedition of 1865*. Vol. 12. The Far West and the Rockies Historical Series. L. R. Hafen and A. W. Hafen (eds.), 60–91. Glendale, CA: The Arthur H. Clark Company.

Cooke, P. S. G. 1857. *Scenes and Adventures in the Army*. Philadelphia: Lindsay and Blakiston.

Cooper, C. L. 1940. Eighty-six Hours without Water on the Texas Plains. *The Southwestern Historical Quarterly* 43, no.3: 360–64.

Coues, E., ed. 1898. *The Journal of Jacob Fowler*. New York: Francis P. Harper.

DeSmet, P. J. 1972. *Western Missions and Missionaries: A Series of Letters*. W. L. Davis, SJ (ed.). Shannon, Ireland: Irish University Press.

Estep, R., ed. 1960–61. Lieutenant William E. Burnet Letters. *The Chronicles of Oklahoma* 38, no. 3: 274–309; no.4: 369–96; 39, no.1: 15–41.

Foreman, G., ed. 1941. *A Pathfinder in the Southwest: The Itinerary of Lieutenant A. W. Whipple during His Explorations for a Railway Route from Fort Smith to Los Angeles in the Years 1853 and 1854*. Norman: University of Oklahoma Press.

Goodwin, C. G., ed. 1970. The Letters of Private Milton Spencer, 1862–1865: A Soldier's View of Military Life on the Northern Plains. *North Dakota History* 37, no. 4: 233–69.

Gray, A. B. 1963. *Survey of a Route on the 32nd Parallel for the Texas Western Railroad, 1854*. L. R. Bailey (ed.). Los Angeles: Westernlore Press.

Hafen, L. R. (ed.). 1942. *Overland Routes to the Gold Fields, 1859 from Contemporary Diaries*. The Southwest Historical Series, Volume 11. Glendale, CA: The Arthur H. Clark Company.

Hafen, L. R., and A. W. Hafen (eds.). 1959. *Relations with the Indians of the Plains, 1857–1861*. The Far West and the Rockies Historical Series, Volume 9. Glendale, CA: The Arthur H. Clark Company.

Hollon, W. E. 1975. *The Great American Desert Then and Now*. Lincoln: University of Nebraska Press.

Hudson, J. C. 1996. The Geographer's Great Plains. In *Occasional Publications in Geography*. Vol. 1. Kansas State University, Number 1, pp. 1–10.

Hughes, J. T. 1848. *Doniphan's Expedition: Containing an Account of the Conquest of New Mexico*. Cincinnati, OH: J. A. and U. P. James.

Jackson, D., ed. 1966. *The Journals of Zebulon Montgomery Pike*. 2 Vols. Norman: University of Oklahoma Press.

Jackson, R. H. 1975. Mormon Perception and Settlement of the Great Plains. In *Images of the Plains, the Role of Human Nature in Settlement*. B. W. Blouet and M. P. Lawson (eds.), 137–47. Lincoln: University of Nebraska Press.

Karl, T. R., and R. W. Knight. 1985. *Atlas of Monthly Palmer Hydrological Drought Indices (1895–1930 and 1931–1983) for the Contiguous United States*. Historical Climatology Series 3-6 and 3-7. Asheville, N.C.: National Climatic Data Center and National Oceanic and Atmospheric Administration.

King, J. H. 1940. Eighty-six Hours without Water on the Texas Plains. W. C. Nunn (ed.). *The Southwestern Historical Quarterly* 43, number 3: 356–60.

Marcy, R. B. 1866. *Thirty Years of Army Life on the Border*. New York: Harper and Brothers.

Mattison, R. H., ed. 1969. The Fisk Expedition of 1864: The Diary of William Larned. *North Dakota History* 36, no. 3: 209–74.

Meko, D. M. 1992. Dendroclimatic Evidence from the Great Plains of the United States. In *Climate Since A.D. 1500*. R. S. Bradley and P. D. Jones (eds.), 312–20. New York: Routledge Press.

Miles, N. A. 1969. *Personal Recollections and Observations of General Nelson A. Miles*. New York: Da Capo Press.

Mock, C. J. 1991. Drought and Precipitation Fluctuation in the Great Plains During the Late Nineteenth Century. *Great Plains Research* 1, no. 1: 26–57.

Moore, C. T. 1999. Nineteenth Century Processes Contributing to the Treelessness of the Great Plains. *Forum of the Association for Arid Lands Studies* 15, no. 1: 87–102.

Myers, F. 1971. *Soldiering in Dakota, Among the Indians, in 1863-4-5*. Freeport, NY: Books for Libraries Press.

Nuttall, T. 1905. *Journal of Travels into the Arkansas Territory, During the Year 1819*. *Early Western Travels*. Vol. 13. R. G. Thwaites (ed.). Cleveland, OH: The Arthur H. Clark Company.

Parkman, F. 1917. *The Oregon Trail: Sketches of Prairie and Rocky Mountain Life*. Boston: Little, Brown.

Peirce, H. A. 1963. *The Second Nebraska's Campaign Against the Sioux*. R. D. Rowen (ed.). *Nebraska History* 44, no. 1: 25–53.

Robinson, J. S. 1932. *A Journal of the Santa Fe Expedition Under Colonel Doniphan*. C. L. Cannon (ed.). Princeton, NJ: Princeton University Press.

Rusling, J. F. 1877. *The Great West and Pacific Coast*. New York: Sheldon and Company.

Saarinen, T. F. 1977. Review of Images of the Plains: The Role of Human Nature in Settlement. *Annals of the Association of American Geographers* 67, no. 2: 295–97.

Sage, R. B. 1958. Scenes in the Rocky Mountains, and in Oregon, California, New Mexico, Texas, and in the Grand Prairies. In *The Far West and the Rockies Historical Series*. Vols. 4 and 5. L. R. Hafen and A. W. Hafen (eds.). Glendale, CA: The Arthur H. Clark Company.

Smith, H. N. 1950. *Virgin Land: The American West as Symbol and Myth*. New York: Random House.

Stahle, D.W., and M. K. Cleaveland. 1988. Texas Drought History Reconstructed and Analyzed from 1698 to 1980. *Journal of Climate* 1, no. 1: 59-74.

Stockton, C. W., and D. M. Meko. 1983. Drought Recurrence in the Great Plains as Reconstructed from Long-Term Tree-Ring Records. *Journal of Climate and Meteorology* 22, no. 1: 17–29.

Stuart, J. E. B. 1959. Journal. In *Relations with the Indians of the Plains*. The Far West and the Rockies Historical Series. Vol. 9. L. R. Hafen and A. W. Hafen (eds.), 217–44. Glendale, CA: The Arthur H. Clark Company.

Thwaites, R. G., ed. 1904. *Original Journals of the Lewis and Clark Expedition, 1804–1806*. 7 Vols. New York: Dodd, Mead.

Thwaites, R. G. (ed.). 1905. An Account of an Expedition from Pittsburgh to the Rocky Mountains, Performed in the years 1819, 1820; Under the Command of Major

Stephen Long. *Early Western Travels,* Vol. 15. Cleveland, OH: The Arthur H. Clark Company.

Townsend, J. K. 1905. Narrative of a Journey Across the Rocky Mountains, to the Columbia River, and a Visit to the Sandwich Islands, Chile. In *Early Western Travels.* R. G. Thwaites, ed., 111–369. Cleveland, OH: The Arthur H. Clark Company.

Walker, S. 1961. *Powder River Campaigns and Sawyers Expedition of 1865.* The Far West and the Rockies Historical Series. Vol. 12. L. R. Hafen and A. W. Hafen (eds.), 92–100. Glendale, CA: The Arthur H. Clark Company.

Warren, G. K. 1856. *Explorations in the Dacota Country, in the Year 1855.* U.S. Congress, Senate Executive Document 76, 34th Cong., 1st Sess. Washington, DC: A. O. P. Nicholson.

Warren, G. K. 1859. *Explorations in Nebraska.* U.S. Senate Executive Document 1, 35th Cong., 2nd Sess. Washington, DC: GPO.

Webb, W. P. 1931. *The Great Plains.* New York: Grosset and Dunlap.

Williams, B. J., ed. 1969. Overland to the Gold Fields of California in 1850: The Journal of Calvin Taylor. *Nebraska History* 50, no. 2: 125–49.

Three Culture Meandering Across the Plains: An Enduring Problematic for Water Politics

John Anderson

Water's precious place in the lives of people living on the Great Plains cannot be doubted.

In fact, the further west of the magical 98th Meridian one goes, the more water becomes vital to agriculture and the very existence of human communities (Foss, 1960; Webb, 1957). This meridian is the imaginary line that marks the beginning of what settlers and explorers used to call the Great American Desert. More recently, this region has just been typified as the "Empty Quarter" (Harrigan, 1994). As the name implies, this part of North America is sparsely populated because of the scarcity of water. Like most scarce resources, water has brought continual conflict. Sharing water has always been difficult, and has left a long history of water theft, wars, and legislative battles in the arid and semi-arid places of the North American continent. It is a truism that water issues rest at the heart of most public debates that arise from the Great Plains.

In the future, greater pressures to use the water found on and under the Plains will bring more conflict. As most observers have witnessed, that conflict will surely generate more and more court rulings and may even generate the outright regulation of water usage (Longo, 1989). Of course, regulatory activities bring conflict (Lowi, 1964). These conflicts tend to create a spiraling effect; as soon as a court, or court-appointed special master writes a regulatory ruling, political lobbying, counter-suits, and tactics of noncompliance ensue. Similar reactions are guaranteed whenever a governmental agency or legislative body writes a regulatory rule. Conflict avoidance is the kind of medicine most reasonable people prefer and this has been the purpose of the prescriptions of nearly every academic observer—avoid conflict through the employment of some special tonic.

Among the most popular remedies, one consistently comes to the front. That oft-prescribed remedy is "culture." Pundits and academics alike have held that some element of culture might be used as a salve that will help prevent the chafing that water battles generally bring to communities and the region. The prescription is simple. It is argued that plains people have a culture based on shared values and common ways of thinking that can be used to settle problems (mostly on their own without outside interference or politics). It has been suggested that a shared culture, or common "identity," has virtually unlimited possibilities for creating understanding (Brooks, 1997). Other scholars have gone so far as to suggest that we change the culture by restructuring it to improve how we decide what happens on the Great Plains (Worster, 1985). Either approach is problematic.

Using a totalized concept like culture as a cure-all fails us in two ways: it neither represents reality well nor shows how we might proceed effectively as a group of people who want to solve our pressing problems. In this chapter, the problems with this type of approach are explored. Further, I argue that any effective approach must tackle the thorny but intractable problem of interests to mitigate its possible negative effects. In effect, the argument is that politics are inevitable but need not be so problematic.

WHAT IS CULTURE?

Understanding the nature of the word *culture* and how it is used is a place to begin. It is a good place to begin because it leads us to an answer to the most important question: Is culture an adequate concept? Answers could take a few directions. Cultural prescriptions might work. They might miss the mark because the term is just too abstract. Perhaps cultures like the putative culture of the Great Plains of North America are too complex and expansive to define (Wuthnow, Hunter, Bergesen, and Kurzweil, 1984). Or maybe we have not done enough work to know how to define culture. Or maybe we are barking up the wrong tree.

Whatever its problems, we know that we cannot say culture is unimportant. Too many instances of the term's importance suggest it has a place in the lexicon of politics, policy, and the study of human society. Besides, knowing something about human cultures is critical to understanding the general human condition (as described by Hannah Arendt, 1958) and it is too important to just toss it. Cultural influences on political machinations have been significant, no matter where they occur. Islamic culture is vitally linked to the politics of Iran. Similarly, it is easy to contend that doing business in Japan requires some understanding of the Japanese culture. Closer to home we find it is quite important to understand that tribal governance and the social problems of Native American reservation life are wrapped in the webs of culture, both aboriginal and American. Even understanding the politics of grass in the arid West is subject to knowing something about cultural ways, or the social environment, of decision makers in these remote places of the continent (Foss, 1960).

We know that culture is a part of many kinds of analysis in the social sciences (sociology, geography, etc.). For the most part culture is used to delineate a very general abstract area for studies that lie outside the formal set of rules that constrain our actions. A touching and insightful book about a small town in eastern Washington illustrates that the realm of culture is important especially when it comes to the social networks, norms, and conventions of small communities (Dillman and Allen, 1994).

> The community norms must be followed. Leaders who fail to follow a norm often find themselves relinquishing their position or adapting a broader community scope when deciding on community problems. The sanctioning process is public and is facilitated by the open, informal meeting held after the formal city council meeting. As in the bank, the norm in Bremer is that politics are open to the public. It is also necessary to remain public, as one community club member explained: "If we weren't public and open about what we do then rumors would get started. Once that happened it is hard to get everybody together again." (120)

Although it may not be readily apparent, the informal ways of doing described by these authors, Dillman and Allen, are part of what makes a culture. In this way, peoples' thinking helps form their approach to life and their shared approach to living. This constraining influence of human thinking is part and parcel of culture. We also know that people in small communities (as well as larger communities) can be actuated as well as constrained by cultural factors.

Culture includes much more than a way of thinking. Culture also includes a way of acting; it is linked to behavior. A famous work in the realm of political and social thought, *Democracy in America* by Alexis de Tocqueville (1981), was written as a description of the American democracy as it existed in the early 1830s. The author wrote that he wanted to study democracy in the United States at that time because of the specialness of American customs (which we should understand as culturally actuated and constrained behaviors).

> The customs of the Americans of the United States are, then, the peculiar cause which renders that people the only one of the American nations that is able to support a democratic government; and it is the influence of customs that produces the different degrees of order and prosperity which may be distinguished in the several Anglo-American democracies. (193)

Those customary behaviors that are motivated, justified, and explained by our language are integral parts of our culture. This is especially true of our political culture where we name our behaviors, like voting, because they represent special actions we value and wish others to value.

A closer look at political culture reveals a little more about the term. Gabriel Almond and Sydney Verba produced the seminal work in the subfield of political culture that described culture as the linkage between micropolitics—writ personal and psychological—and macropolitics, the study of institutions, nations, and other organizations acting on the international and intranational levels (Almond and Verba, 1963). Almond and Verba, like Allen and Dillman, gave us a conceptual place for political culture but they did not make culture an objective reality like that embodied in the institutions and organizations found at the macrolevel of analysis, or in the person herself, who is both subject and object, at the microlevel of analysis.

In another well-known work on political culture, Daniel Elazar (1966) tried to build on the work of Almond and Verba but at the intranational (or state) level. When he addressed culture in his analysis, Elazar only used the term as a referent for peoples' thinking and attendant actions. Elazar did add to our understanding somewhat by suggesting that some forms of thought came to people from traditions they had experienced; in other words our ancestors gave us traditions, norms, and other ways of approaching life.

What stands out in the treatment of culture by both Almond and Verba and Elazar is that they accomplished much by introducing culture but they also did little to clearly define it. They simply placed it outside institutional understandings, and somewhere in the world we learn from others. This is certainly a weakness. One cannot prescribe a very general phenomenon when the problems are specific.

WHAT BUILDS CULTURE?

Beyond the general arena of culture is a large penumbra, to borrow a term from astronomy, of social actions and language that builds and supports the culture in which we live. Scholars like Robert Putnam have worked to describe and explain parts of the forms of human activity that lend to our political world. Putnam explored how people in Italy have developed "social capital." According to Putnam's (1993) research, those areas of Italy with a strong republican tradition and a history of working together in secondary groups, like clubs and even soccer clubs, are regions with governments that work well and the people in those regions are more prosperous as well. Apparently, their culture is strong and supportive of cooperative efforts that connect culture directly to institutional mechanisms. Putnam also found that a different culture existed in the southern reaches of Italy. Putnam (1993) explored parts of Italian political and social culture to see how it affected economic and political life there and he added to our understanding of culture by showing how it was linked to institutional actions.

Recently, another set of scholars began to explore an idea they call identity. Say what you will about identity and the scholarship around the term, there can be little doubt that it is part of the same idea other scholars call culture (Wuthnow et al., 1984). Identity and culture are similar simply because persons' shared identities are expressed through their customs, language, thoughts, and actions, all of which have been defined as basic components of culture. Identity is also associated with the kinds of actions people take in concert. I am a professor, and professors attend graduation ceremonies, therefore I normally attend those ceremonies. There is most assuredly a much larger set of examples I could offer to show how identity works to bring common action, or what we can call collective action. Knowing whether residents of the Great Plains see themselves as such can add a little more specificity to the talk they talk when it comes to culture in the Great Plains.

In sum, it seems that culture, especially in the political world, is an abstraction that covers an enormous conceptual terrain but it does hold some meaning. Culture is linked to actions and thoughts. In this way, the general phenomenon we call culture is supportive of kinds of political and social action. In fact, certain components in the cultural realm may be necessary for certain kinds of governance, such as democracy in America or Italy. One reason parts of a culture can be used to support democracy is that these cultural components enhance democratic practices; but, culture can also constrain action by making certain activities taboo or by establishing some form of social sanction. Furthermore, we may find what Putnam calls "negative social capital" (Putnam, 1993). In these cases, people know how to work together and do so to the detriment of others. Consider gang interactions and actions as good examples of negative social capital and another kind of negative cultural activity. Regardless of how parts of any culture are used, as either a prompt or a constraint, our most truthful observation ought to be that the term culture represents one of the most general forms of abstraction people use to talk about themselves.

Culture, as a term, lacks specificity and it tends to simplify the very sources of conflict in difficult political-policy matters, such as the ones generated by scarce water on the Great Plains. In particular, the idea of a common culture or shared "ethic" does little to solve the real problems that lie at the heart of water issues. In fact, general forms of abstraction that imply shared sympathies, whether they are Worster's environmental ethic or the use of some conception of common identity, attempt to do the impossible by blurring real differences.

WHAT ABOUT PLURALISM?

A careful reading of political matters will show that pluralism permeates the social and political setting wherever one looks in America (Dewey, 1954). Ian Frazier's book, *Great Plains* (1988), gives a portrait of this vast region, which suggests that differences are the norm. For instance, consider Native American experiences in which a set of diverse tribal experiences should make any observer ask whether the culture was Plains or Great Basin, not whether it was a Native American culture. Or better yet, observers should ask whether the Native Americans were Ponca, Dakota, Lakota, Shoshone, Kiowa, Northern Cheyenne, Blackfoot, Pawnee, or from some other important tribal culture. Similarly, visiting the plains (or reading Frazier's book) should lead one to wonder whether the people in any plains setting were farmers or town people? from the Red River Valley or from the Platte River Valley? from a town whose history is linked to a Czech, German, Irish, or Lithuanian heritage? Pluralism is the rule on the Great Plains.

A quick examination of scholarship shows that a plurality of approaches to water, whether on the Great Plains or not, is the rule in water debates, too. Whether water is abundant or scarce, each situation and way of thinking differs.

On the Great Plains, the debate started by Karl and Deborah Popper over the reintroduction of a bison economy is illustrative of the diversity that exists in this region. Whether you looked East or West or North or South made a difference in the views one heard expressed about re-introducing buffalo in large numbers. In the West, there was resentment toward Easterners bringing ideas and suggestions about how life in the West should look. One Wyoming official put it best when he said, "Perhaps people in the Plains don't see much wisdom coming from Easterners who have screwed up their states so much" (Maxfield, 1991, 3). On the other hand, one Minnesotan thought it was a great idea, simply because he already had a real interest in buffalo—this Minnesotan owned a buffalo operation. Other prairie residents were quick to respond that the Poppers' thesis painted the economic viability of the region with a brush stroke that was too broad. Although parts of the Great Plains suffered through droughts, others in different parts survived quite well. Certainly, droughts can cause problems in places like western Kansas but seldom affect farmers in the Red River Valley of North Dakota and south central Canada to any great degree. At least, we can suggest residents of the Great Plains differ significantly in their economic interests and that affects their identity and general culture.

When we look at water and culture across a wider spectrum we also find variation on the Great Plains; this plurality of differences generally brings conflict, not accord (Donahue and Johnston, 1998).

For instance, the Hopi Indians and Peabody Mining have been locked in a battle over their opposing interests—the use of well water. The Hopi oppose Peabody's use of well water to slurry coal out of New Mexico. Peabody as an industrial organization and the Hopi as aboriginal people have varying interests that go far beyond cultural understanding and that drive each to stand opposed to the other (Whitely and Masayesva, 1998). At the very least we can argue that their ways of thinking cannot be separated from their interests.

The Katy Prairie near Houston, Texas, is another example of a plurality of interests when it comes to water and land use. We might wish that a land and water ethic would replace interests there, but that is a chimera. When Audubon Society members saw the Katy Prairie they plainly saw a wild place that needed wetlands for birds. When developers saw the prairie they saw a place for a new airport and the income its development would bring to people (Berryhill, 1994). The interests, economic and recreational, of folks involved in that debate are really no different from the interests we possess and the debates those interests cause each of us to experience.

Our own place in life can simultaneously give us some kind common identity and divide us. This is even true on the Great Plains where political movements have proliferated. In an article on this subject, Mildred Schwartz (1997) argued that residents had much to share but admitted that age, sex, race and ethnicity, language, religion, and geography all play a role in forming approaches to political argument among people in this region. Although these influences on identity make us similar in one way, they make us different in other ways. So, as we find continuities on some matters, we also find great differences among Plains residents. Schwartz argued that common conceptions of problems and viable solutions inevitably lead to movements that cross borders. This conclusion certainly must be true, but we can argue she missed the particularities that strain group movements. We should note that leaders' inability to continually mobilize movement members is an indication that divisions among social movement members make cohesion difficult over a long period of time. Else, why are most of the members of the Grange, the Nonpartisan League, and the Populist Party no longer with us? Of course, the times have changed the people but their natural differences, in terms of interests, caused changes in behavior (group membership), too.

Seeing an abundance of varying experiences might be what motivates scholars and pundits to look for a common culture, a common identity, or a common ethic. Reasoning could drive a person to look for some idea of what could be held in common. On the other hand, the facts of living in a pluralistic setting in which people hold certain interests that drive them to political action in association with other people need not be seen as problematic.

INTERESTS AND GOOD POLICY

Aside from missing the importance of a plurality of peoples on the plains, we can see a plurality of differences. Enter the political philosophy of James Madison—a pluralist. Madison was grounded in the classical liberal tradition began by Thomas Hobbes and John Locke. Madison is best known for being the father of the Constitution and for later becoming the president of the United States; he is also noted for his writings in a series of pamphlets designed to persuade New Yorkers to ratify the Constitution. The pamphlets are simply called the *Federalist Papers* and Madison's #10 is the best known of them all. This piece of political writing explains how "interest" works to divide people into groups (or factions, as Madison termed them).

Madison understood interest as being a basic part of human nature and the human condition. In Madison's words, people are united into factious groups "by some common passion, or of *interest,* adverse to the rights of other citizens, or to the permanent and aggregate interests of the community" (emphasis mine) (Madison, 1982, 43). This quality he finds to reside in humans because they live on the earth and must work to gain wealth for the purposes of satisfying their needs and wants. Of course, wants may be manufactured but individual passions and interest in wealth are primary and not crude. Given a somewhat limited supply of wealth, there are naturally some differences that will bring conflict regardless of the shared understandings held by men and women. However, Madison saw more. Madison believed that it was also our individual differences that created our differences on matters of policy and each person's view of the community weal. In *Federalist #10,* he continued to argue, "The diversity in the faculties of men from which the rights of property originate, is not less an insuperable obstacle to a uniformity of interests" (43–4). With so much diversity among us, even within the small states immediately after the Revolution, how could the founders, or we today, look to our commonness with any assurance that general understandings will permit us to get along and get along well? This seems to be the crux of the matter when it comes to governance but, in particular, it is the problem we face when we attempt to regulate who uses water.

To control the causes of violence, or cheating, Madison placed government at the national locus of control with some powers to resolve conflicts around economic interests. He reasoned that a republic that was far-reaching could never come under the control of a single interest. Of course, without a single view as to the right action we are sometimes left dangling over difficult issues. At the same time, though, we know that no small minority (or even a majority in a small setting) will be able to dominate another group. We must keep the institutions intact that provide that kind of assurance. There are few guarantees that disputes over water rights can be easily resolved without governmental structures to ensure the rights of minorities or to ensure that the common interests, whatever they may be, are served. We cannot assume we will suddenly be able to form common conceptions and resolve all our disputes; we should never be tempted to believe that we can move past the understanding that people are actuated by their passions and interest.

Certainly, from an evaluative perspective there is some reason to doubt that using culture, the formation of a common identity, or the sudden appearance of a new ethic will be workable for a people who believe in protecting the rights of all. These approaches fail as accurate descriptors and suggest an approach that is fraught with failings, especially if our normative approach includes any concerns for the rights of minorities and the more general public interest. In other words, the problem of interests is intractable—no cure can be found for it.

WHAT ARE WE TO DO?

Disillusionment with employing a common culture to settle conflicts does not mean some portion of our shared understandings cannot be used to help solve water problems on the Great Plains, or elsewhere. It also does not mean that the current economic ethic we use to address our use of water need be reified. We can continue to better understand the meanings of our relationships with others and the world in which we live one small step at a time. Within our governmental institutions lie certain protections against the possible ills of a common approach. Beyond that, certain tools may be of great benefit.

First, there is a burgeoning field of research that suggests the people of the Great Plains possess an enormous reservoir of social capital. This is shown by recent research that rates the states of the upper Midwest above all other states in terms of social capital (Putnam, 2000). Data collected in Nebraska affirms that associational life is strong in most of the communities in the state. This is especially true of towns with populations between 1,000 and 5,000 (Anderson, 2000).

Although it may be argued that social capital is a part of that space between the person and institutions Almond and Verba (1963) called culture, social capital is not culture. In its earliest formulation, social capital was used by John Dewey to refer to actual social institutions students might access once they were ready to act in the social world (Dewey, 1990, 111). Dewey wrote about a form of capital (the basic material necessary to develop something through action) found in the social realm that makes social action possible (111). One could say that social capital is a part of what makes it possible for people to cooperate and accommodate their differences on many but not all kinds of issues. Here, rather than an idea (such as culture) we have the places and relationships where the kinds of practices people employ within the constraints of their environment are what matter.

Dewey's social capital may be much like the phenomena Alexis de Tocqueville witnessed in his travels across the American Republic in the early 1830s. De Tocqueville witnessed a level of cooperation and engagement in local political matters among Americans he thought was unparalleled. This French scholar suggested it was mostly motivated by something the Americans called "self-interest rightly understood" (415). Self-interest rightly understood simply means people consider their neighbors when they consider their own plight. It is a way of thinking that draws the individual away from a simple definition of self to include others, even all others. This can be important when water conflicts are conceptualized because think-ing about your neighbor widens the sphere of interest. It does not suggest that we can bring farmers from the Central Platte together with farmers from the Upper North Platte and expect them to share views and interests. This suggestion is similar to Donald Worster's (1985) argument for watershed-defined communities because it rec-ognizes the importance of face-to-face understanding. Nor can we place town people from Kearney across the table from farmers who live on the Platte River near Columbus and expect them to have enough self-interest rightly understood to enter into a conversation that will settle matters easily.

Work needs to be done to see how each person shares some inter-est with others. The work is necessary because shared interests are not readily apparent to most people when they do not have opportu-nities to talk to others with similar problems. In fact, people probably need to come directly to the table with each other before any practical understandings can be sought. When federal and state agencies, en-vironmental groups, and agricultural interest groups began to work on a memorandum of agreement to govern decisions about the use of Platte River water, little was accomplished until all came to the table and had equal voice (Ring, 1999).

This brings us to the place where people can understand that the difficulty of settling matters must be anticipated and viewed as nor-mal and good. It may take longer to form commonly accepted means for living with water problems than to go to court, as hard as that may be to believe. Any reasonable model for addressing differences will require participants to work to build necessary and limited under-standings—in a sense provisional truths. This work is always going to be difficult because diverse interests and understandings are the norm. The process of working through a problem concerning water will take time. Again, the Platte River Agreements took many years to reach—well over five tough years of talking (Ring, 1999).

Finally, if we are to bring normative considerations to the table, some form of deliberative processing should be considered. As re-searchers interested in deliberative polling begin to refine their tech-niques, more and more people are finding it to be a preferable ap-proach to reaching common understandings. Using methods similar to focus group techniques brings people together and making them consider others' plight is worth the effort. When you consider your neighbor's problems in person, you have to begin to assess these problems in a different light just as you hope your neighbor at the table will think of your problems differently. Although this sounds almost like the golden rule, it is a very practical understanding that works among groups of people who believe they had better work together. Here, the force of governmental rule or the threat of a probable suit may work to compel compromise, which is not a bad word when people are trying to settle conflict.

By contrast, some have criticized deliberative activities because organizers often fail to readily permit or recruit all of the interested parties to the table (Sanders, 1997). It is difficult to interrupt people's lives to have another meeting, but the social capital and experiences that come with it may work to help surmount the problems. While forming the agreement governing water-use decisions on the Platte River, a handful of the people at the table simply stopped participating for a variety of reasons that are quite common. Some stopped because they thought they could not be successful and others stopped because they transferred to new jobs (Ring, 1999). The ones who continued to attend probably knew about meetings and readily lent assistance to the business at hand rather than staying at home to watch their favorite television program.

We must also realize that we are not looking for progress. Progress as an ideal can become a huge stone around the necks of people trying to settle conflicts over water. Essentially, we face a problem because one person's progress is not another's. Any idealized term becomes a problem when it cannot be obtained, and this type of problem presents a very difficult obstacle when people need to work together; eventually frustrations set in when people believe they should be moving forward. On the other hand, looking for a series of small wins and celebrating them may be a very good idea. Rather than moving forward, people involved in a conflict over water should be very happy when they can reach some kind of accord that allows them to see the conflict settled without great adversity for any of the participants.

Like a military general in the field, it is good to win a position and if you win enough you may succeed in keeping your army in the field and alive. This is more or less what we should seek with water conflicts. Although the infamous Carol Sheldon of Nebraskans First may worry about his loss of the ability to use as much water as he has in the past, in the long run he would do well to worry about just getting enough water to keep farming. This may not come about if any interest group leaders enter into a courtroom where zero-sum games are played. Mr. Sheldon and other Nebraskans who farm in the Central Platte River Valley may be here for quite some time if they can simply have enough miniwins to keep irrigating their land. Of course, this is good advice for all who enter into conflicts over water on the Great Plains.

REFERENCES

Almond, Gabriel, and Sydney Verba. 1963. *The Civic Culture: Political Attitudes and Democracy in Five Nations.* Boston: Little, Brown.

Anderson, John, L. 2000. Measuring Social Capital in 30 Nebraska Towns. Paper presented at the Northwest Political Science Association Meeting. November. Portland, Oregon.

Arendt, Hannah. 1958. *The Human Condition.* Chicago: University of Chicago Press.

Berryhill, Michel. 1994. Crusade to Save the Katy. *Audubon* (Nov.-Dec.): 122–33.

Brooks, Kevin. 1997. Liberal Education on the Great Plains: American Experiments, Canadian Flirtations, 1930–1950. *Great Plains Quarterly* (Spring): 103–17.

de Tocqueville, Alexis. 1981. Introduction to *Democracy in America* by Thomas Bender. New York: Modern Library.

Dewey, John. 1954. *The Public and Its Problems.* New York: Swallow Press.

Dewey, John. 1990. *The School and Society, and the Child and the Curriculum.* Chicago: The University of Chicago Press. Originally published in 1900.

Dillman, Donald A., and John C. Allen. 1994. *Against All Odds: Rural Community in the Information Age.* Boulder, Colo.: Westview Press.

Donahue, John M., and Barbara Rose Johnston. 1998. *Water, Culture, & Power: Local Struggles in a Global Context.* Washington, D.C.: Island Press.

Elazar, Daniel. 1966. *American Federalism: A View from the States.* New York: Thomas Y. Crowell.

Elazar, Daniel. 1986. *Cities of the Prairie Revisited: The Closing of the Metropolitan Frontier.* Lincoln: University of Nebraska Press.

Foss, Donald. 1960. *Politics and Grass: The Administration of Grazing on the Public Domain.* Seattle: University of Washington Press.

Frazier, Ian. 1988. *Great Plains.* New York: Farrar, Straus & Giroux.

Harrigan, Anthony. 1994. The Great Plains of America: The Therapy of Distance. *Contemporary Review* (Jan.): 9–16.

Harris, Marvin. 1979. *Cultural Materialism: The Struggle for a Science of Culture.* New York: Random House.

Longo, Peter J. 1989. The Constitutionalism and Water Policy of Sporhase Revisited: A West German Alternative. *Environmental Law* 20, no. 4: 417–28.

Lowi, Theodore J. 1964. American Business, Public Policy, Case Studies and Political Theory. *World Politics* 16: 677–715.

Madison, James. 1982. *Federalist #10, The Federalist Papers, by James Madison, Alexander Hamilton, and John Jay.* New York: Bantam Books.

Maxfield, Peter. 1991. *The Alliance (Alliance, Nebraska) Times-Herald,* 16 August.

Putnam, Robert D. 2000. *Bowling Alone: The Collapse and Revival of American Community.* New York: Simon & Schuster.

Putnam, Robert D., with Robert Leonardi and Raffaella Nanetti. 1993. *Making Democracy Work: Civic Traditions Modern Italy.* Princeton, N.J.: Princeton University Press.

Ring, Ray. 1999. Saving the Platte. *High Country News,* 1 February. Online at http://www.hcn.org.

Sanders, Lynn M. 1997. Against Deliberation. *Political Theory* 25, no. 3: 347–77.

Schwartz, Mildred A. 1997. Cross-Border Ties Among Protest Movements: The Great Plains Connection. *Great Plains Quarterly* 17 (Spring): 119–30

Webb, Walter P. 1957. The American West: Perpetual Mirage. *Harper's Magazine,* May, 25–31.

Whitely, Peter, and Vernon Masayesva. 1998. The Use and Abuse of Aquifers: Can the Hopi Indians Survive Multinational Mining? In *Water Culture and Power: Local Struggles in a Global Context.* John M. Donahue and Barbara Rose Johnston (eds.). Washington, D.C.: Island Press.

Worster, Donald. 1985. *Rivers of Empire: Water, Aridity & the Growth of the American West.* New York: Pantheon Books.

Wuthnow, Robert, James Davison Hunter, Albert Bergesen, and Edith Kurzweil. 1984. *Cultural Analysis: The Works of Peter L. Berger, Mary Douglas, Michel Foucault, and Jurgen Habermas.* New York: Routledge Kegan Paul.

Part Two

Realities: Economics, Law, and Politics

Under the Volcano

They live under the volcano   *the peasant's song*
They carried lanterns   *into the storm cellar*
Pulling the wood cart out of the mud *takes a little time*
They saw a vision   *angel blood*
They washed babies   *born in a shallow river*
They emptied the coal bin   *under winter light*
The peasant's song   *on the ice pond*
Under the snow   *the creature awakened*
They were born on a river   *wade in the water*
They gather flowers   *elixir and wool*
Takes a little time   *pulling the wood cart out of*
     the mud

Under winter light   *the earth shall smoulder*
The waterfall   *God's lacerated tongue*
Under the volcano   *snow*

Charles Fort

The Petrified Forest

They drove for days	*into the Petrified Forest*
French poetry	*among thieves and washtubs*
The wing and ash	*flew out of the hollow*
He brushed the cactus	*in a tweed jacket*
The sidewinder	*under the trough*
Tarantula dance	*in the holy desert*
Words spoken by the trees	*to the citizens and thieves*
The snake's trough	*wing and ash*
In the Petrified Forest	*a rumble seat jalopy*
Tweed and majesty	*in the wilderness*
Strong black coffee	*or snake venom?*
In the holy desert	*tarantula dance*
Thundercloud	*God's tambourine*
The citizens and thieves	*had spoken to the trees*

Charles Fort

Four The Use of Equitable Principles to Resolve "New" Western Water Disputes

J. David Aiken

Irrigation accounts for over 90 percent of total western water consumption (Aiken, 1980a). Historically, state water policies in the Great Plains and the West generally have been de facto irrigation policies rather than more holistic water policies. As the competition for water to supply a variety of uses (including environmental uses) intensifies, it is time for water policy makers to consider establishing minimum water-use efficiency standards in order to achieve a fairer distribution of rights to use this important public resource.

Traditionally, water right disputes (typically among appropriators) in the Great Plains and the West have been resolved based on priority: "first in time is first in right." This rule of priority developed from California eighteenth-century mining camp customs in which the first miner to stake (i.e., mark) a mining claim had the first right to work that claim so long as the miner continued to work it. If the miner abandoned the claim, then the claim was available to the next claimant. Miners often diverted and used surface water in working their claims, and the priority rule that applied to the mining claim also applied to the water diverted to work that claim. The California Supreme Court adopted these mining camp customs in initiating and developing the American water law doctrine of prior appropriation.

Prior appropriation has worked well in that it provides clear rules for resolving conflicts to use overappropriated western streams. (An overappropriated stream is one in which too many appropriations have been granted, and junior appropriators are routinely issued closing orders by the state engineer whenever a senior appropriator makes a priority call. On overappropriated streams, junior appropriations do not represent a reliable water supply on average.) However, prior appropriation has its share of critics, contending *inter alia* that it rewards waste (Pring and Tomb, 1979), is inflexible, and gives an unfair advantage to the earliest (i.e., senior) users. (Under the prior appropriation doctrine, the earliest uses have the oldest priority dates, typically the date when the water use was first initiated. These senior appropriators are entitled to administrative protection of their priorities. When senior appropriators do not receive all the water they are entitled to, they make a priority call. This means that the state water administrator, typically called the state engineer, will issue closing orders to upstream junior appropriators, those who have later priority dates, requiring the juniors to stop diverting water until the senior water use has been satisfied.)

Most of these criticisms are valid. Water marketing, which allows new users to buy senior appropriations when unappropriated water is unavailable to supply new uses, provides significant flexibility in accommodating new water uses and new patterns of water use. Moreover, at least some Western states have taken significant legal steps to reduce appropriation waste, although not on a widespread basis. Two cases illustrate this fledgling trend. In a 1990 California decision, the state required the Imperial Irrigation District to line canals and make other system improvements to keep the district's water from flooding irrigation project area lands (225 Cal App3d 548, 275 Cal 250, 1990). In a 1993 Washington case, an irrigation appropriator was allocated a substantially smaller quantity of water based on the amount needed if the irrigator changed to sprinkler irrigation equipment (the most expensive and the most efficient irrigation distribution system) and if conveyance losses from an unlined irrigation canal were reduced from 67 percent to 25 percent (121 Wash2d 459, 852 P2d 1044, 1993). In contrast to the Imperial Irrigation District case, which forces the irrigation district to reduce gross water waste, this case went further and requires the irrigator to meet a very high standard of water use efficiency. Probably this meant that the irrigator reduced the number of acres irrigated rather than upgrade his irrigation system and distribution system to the level of efficiency required to fully utilize his appropriation.

Despite these movements towards requiring greater water-use efficiency, the priority doctrine still gives a significant and perhaps undue advantage to senior appropriators. When there is insufficient water for all it does seem unfair that the earliest users are legally entitled to all the water they can use and all others must share what is left over. This is in contrast to the correlative rights of California groundwater law, under which all users are treated the same and all are entitled to a proportional share of the available supply during shortages (Aiken, 1980b). The nature of water right conflicts is changing from traditionally being principally conflicts among individual (typically irrigation) appropriators to the more recent conflicts involving significant other water interests, including Indian claims or environmental claims. In view of these emerging conflicts, it is appropriate to consider whether priority of appropriation provides too narrow a basis for effective dispute resolution. This is implicitly recognized in that resolution of contemporary western water disputes often will involve a negotiated settlement among the affected parties rather than a simplistic resort to priority (although the threat of the new use asserting a "super" priority obviously affects the negotiations). This chapter 1) explores the water right theories that are more explicitly based on equity rather than simple temporal priority; 2) analyzes the Restatement of Torts (ALI, 1979) as the most comprehensive equity-based system for water right conflict resolution; 3) examines Colorado conjunctive use regulations as one model for dealing with water supply shortages; 4) evaluates the Platte River Cooperative Agreement as a recent model in which endangered species requirements have been superimposed on a highly appropriated (if not overappropriated) stream; and 5) explores how equitable principles might further be applied in resolving contemporary western water right disputes.

"NEW" WESTERN WATER DISPUTES

Traditional western water disputes typically focused on disputes among individual appropriators. Under the appropriation doctrine mentioned earlier, the earliest or "senior" appropriator is entitled to receive the water at the expense of upstream later-in-time or "junior" appropriators. The senior appropriator can issue a priority call against upstream juniors, because their water diversions affect the amount of flow available to the senior. The senior appropriator cannot issue a call against downstream juniors however, because the water has already passed by the senior, and whether the downstream junior uses the water or not does not affect how much streamflow is available to the senior. Thus, priority runs upstream but not downstream. Although the appropriation doctrine has received its share of criticism as a well considered water allocation doctrine, even its critics admit that it does provide a predictable means for resolving the frequent water-use disputes that arise on often overappropriated western streams. However, there are a number of different contemporary water-use disputes for which the priority system of the appropriation doctrine is clearly lacking. These disputes include assertion of dormant Indian or federal water right claims, recognizing the effect of withdrawing tributary groundwater on senior surface appropriators, and superimposing environmental demands on overappropriated streams (Dunning, 1987).

Dormant Indian or Federal Claims

Under the reservation doctrine, Indian tribes are entitled to appropriate water for reservation use with a priority date that relates back to the date the reservation was established. This is in contrast to the traditional appropriation doctrine that priority dates are based on when the water was first used. Under the reservation doctrine, a hypothetical Indian tribe could initiate a water use in 2000 (the date of first water use) that would have a priority date of 1860 (the date the Indian reservation was established). Clearly, this tribal water use would displace many if not all previously senior appropriators on the stream.

Often Indian reserved rights conflicts arise when a new reservoir is proposed to dam a stream that is subject to Indian reserved water rights. The typical pattern is that the parties will negotiate what share of the water project's supply will be allocated to tribal uses, and what supply remains for those seeking project construction. If the parties can negotiate a division of the water, the Indian reserved rights issues are resolved and the water project can be implemented.

Similarly, federal officials may claim federal reserved water rights for use in federal parks, forests, and monuments. The priority date of the federal reserved water right would date back to the time the federal park, forest, and so on were established. These dates are generally much more recent than Indian reservations and, consequently, are somewhat less disruptive. However, in either case, proceeding simply based on priority (and the new adjusted priority schedule) may work a severe injustice to previously senior appropriators whose uses have been displaced.

Tributary Groundwater

Although it is widely recognized that groundwater is the source of a stream's base flow, it has proven much more difficult to legally coordinate the rights of use for interrelated surface water and groundwater supplies. Most Western states apply the doctrine of prior appropriation to surface water and groundwater. However, few Western states have successfully integrated the administration of interrelated surface and groundwater rights. Further, the simple administration of priority in an interrelated surface and groundwater system will place the typically junior groundwater appropriators at a significant and probably unfair legal disadvantage. A simple application of the priority doctrine would result in junior groundwater appropriators being issued closing orders to stop withdrawals in response to a senior priority call unless the junior groundwater appropriator can demonstrate that the streamflow depletion effect would occur after the senior needed the water. On the other hand, failing to restrict the groundwater withdrawals of junior appropriators could in the long run result in no streamflow, if groundwater levels fell below the stream. Colorado has made the most progress in integrating the competing uses of surface and groundwater (MacDonnell, 1988), as discussed later in this chapter.

Environmental Disputes

The most recent "new" western water disputes involve protection of endangered species. Under the federal Endangered Species Act, western water appropriators may be required to reduce their water use in order to avoid harming endangered species. Federal courts have ruled that private water appropriators are not exempt from the species protection requirements of the Endangered Species Act (Aiken, 1999). In particular, priority of appropriation is legally irrelevant in endangered species proceedings. If endangered species disputes were resolved based on priority, and the endangered species priority was dated from when the species was officially designated as threatened or endangered, the endangered species would rarely receive any water. Conversely, if endangered species receive the absolute highest priority (the effect of current law), existing appropriators have essentially lost their water rights to the extent those water rights are needed for endangered species protection. Consequently, states, water users, and federal environmental managers have begun negotiating settlements or compromises seeking to accommodate the competing interests of appropriators and endangered species. The Platte River Cooperative Agreement, discussed later in this chapter, is a recent example of such a compromise.

In all three of these situations, the policymaker is faced with difficult choices. When water appropriations were made in compliance with state law in the nineteenth century, it was impossible to predict that federal courts would subsequently allocate water to Indian tribes that would displace existing appropriators. Even when the Indian reserved rights doctrine was legally established, it was difficult for policymakers to discern how those rights would be quantified and established in the future, even if their retroactive nature was fully appreciated. When state water appropriations were made in the nineteenth and early twentieth centuries, it was difficult to appreciate that deep-well technology would develop and subsequently would facilitate economic exploitation of groundwater resources, the use of which could over time reduce streamflow. When well drilling and pump technology facilitated groundwater development, it would be decades before any significant streamflow depletion would become manifest. Similarly, the development of analytical techniques to realistically quantify stream-aquifer interaction to make the deliberate management of integrated surface and groundwater supplies would lag the overappropriation of those supplies by decades, resulting in few or no good management options. When western water supplies were developed in the nineteenth and twentieth centuries, no one could reasonably foresee that future generations would place the habitat requirements of endangered wildlife species above legally established and recognized water uses for agricultural, industrial, municipal, and domestic purposes.

In these difficult water management situations, traditional water rights doctrines provide few useful tools. The "California approach" to dealing with water shortages of building new reservoirs to impound additional supplies has fallen out of favor (Reisner, 1986). Instead, water marketing, the selling of existing appropriations for new uses, has become the water management tool of choice to reallocate limited water supplies (Reisner and Bates, 1990). And water marketing can be an effective tool for dealing with many water supply management challenges. However, the availability of water marketing reflects the inadequacy of appropriation to deal with what may be characterized fairly as the water law system failures of dormant rights, tributary groundwater, and endangered species conflicts. In cases in which water marketing cannot reasonably address the water supply imbalance that the new or unaccounted for uses now impose on the system, supplementing prior appropriation with equitable principles may help reach a fairer result. Some sort of compromise is required, and a review of equitable principles may suggest how these necessary compromises may be developed.

EQUITABLE ELEMENTS IN WATER RIGHTS DISPUTE RESOLUTION

Although equity has a very limited role in resolving western water right disputes, eastern water law is based primarily upon equitable principles. The delineation of how equitable principles could be used to resolve water disputes has been most fully developed in the Second Restatement of Torts (ALI, 1979). Equitable principles are also more significant in resolving interstate water disputes under the U.S. Supreme Court doctrine of equitable apportionment than they are in prior appropriation.

Eastern Water Jurisprudence

Riparian Theory
Riparian rights are based upon owning land bordering a stream or other water body. The current American version of riparian rights is the reasonable use doctrine, although states are beginning to follow the Restatement of Torts rule (ALI, 1979) as discussed in the next section. Under the reasonable use doctrine withdrawals of water may be made for use on riparian land (i.e., within the watershed) (Tarlock, 1999). When disputes arise between riparians competing for the same water, courts attempted to resolve the dispute based on each party having an equal right to use the water. The court will consider

1) the nature of the stream (i.e., flow quantity), 2) the types of competing water uses and their impact on the stream, and 3) the balance of benefits of the proposed (i.e., the new) use with harm to other riparians (Tarlock, 1999).

Restatement of Torts

The Restatement of Torts (ALI, 1979) provides the most thoughtful and consistent treatment of how the claims of competing water users should be balanced, and extends riparian right theory beyond the reasonable use rule. It should be noted that the series is a significant attempt to provide the best legal thinking on common law topics to the legal profession. For nonattorneys, state judges actually make or determine the common law through court decisions, using prior court decisions as legal precedent. The series has exerted a significant influence on American state judges in making common law decisions, reflecting the best current legal thinking on virtually every common law topic.

Riparian rights are common law water rights, rather than rights based upon state statutes (such as Western state prior appropriation statutes). Consequently, riparian rights are addressed in the Restatement of Torts, and the Restatement water conflict provisions have been a major influence in the development of current eastern riparian rights legal theory. Under the Restatement doctrine a riparian is liable for making an unreasonable use of water harming another riparian's water use or land (ALI, 1979). The issue, then, under the Restatement is what constitutes an unreasonable use. Several factors are to be considered in a judicial determination of whether the complained riparian use is unreasonable:

1. the purpose of the interfering use
2. the suitability of the interfering use to the watercourse
3. the economic value of the interfering use
4. the social value of the interfering use
5. the extent and amount of harm it causes
6. the practicality of avoiding the harm by adjusting the use or method of use of one riparian proprietor or the other
7. the practicality of adjusting the quantity of water used by each proprietor
8. the protection of existing values of water uses, land, investments, and enterprises
9. the justice of requiring the user causing the harm to bear the loss (ALI, 1979)

These criteria and their potential application to new western water disputes are discussed next.

The general effect of the Restatement criteria is to guide the court in balancing the equities between competing surface water users. The Restatement rule provides a comprehensive list of factors to be judicially balanced in resolving water-use conflicts, in an attempt to accommodate new "progressive" water uses but not at the expense of relatively efficient existing uses.

Western Water Jurisprudence

Priority

Under the prior appropriation doctrine, surface water conflicts are resolved based on priority—first in time is first in right. Thus, the earlier user wins, so long as that user can prove the priority (i.e., the earlier date). The priority doctrine has been criticized as freezing water uses in a pattern of economically inefficient use and providing little flexibility to meet emerging water needs. However, the widespread adoption of water marketing in most Western states significantly ameliorates this concern and provides a significant degree of flexibility in allowing water-use patterns to change in order to accommodate emerging economic, social, and environmental priorities. This flexibility is provided while still providing appropriators with significant water right security: existing water uses may be maintained, but water users may sell their rights when a buyer offers the right price.

Physical Solution

Under the theory of the physical solution, courts may require senior appropriators to accept a substitute water supply from junior appropriators that enables the junior appropriators to use water that would otherwise be used by the senior appropriator. In other words, the junior appropriators can use the senior appropriator's water, but only after the junior appropriator provides a substitute water supply to the senior appropriator. An example would be a downstream senior appropriator who has access to groundwater and surface water, and the upstream junior appropriator has access only to surface water. Under the theory of the physical solution, if the junior appropriator installed a well to supply the senior appropriator, the senior appropriator could not complain if the junior appropriator used the senior appropriator' s share of surface water. In essence, the senior appropriator is required to sell his or her water to the junior appropriator for the well (Dunning, 1986).

The primary relevance of the physical solution here is that the theory challenges junior appropriators to use considerable ingenuity to provide a physical solution to a water problem that enables both senior and junior appropriators to continue using water shortages. Under the physical solution doctrine, courts will reward this ingenuity with water even where a water market would otherwise not operate.

Correlative Rights

The correlative rights doctrine is worth briefly noting as a major contrast to the priority doctrine. The distinctive feature of correlative rights is its sharing approach to groundwater shortages. When water shortages occur, all users must make proportional reductions in withdrawals so that all users receive some water (Aiken, 1980b). This pro rata sharing principle is in stark contrast to the priority doctrine, in which senior appropriators receive a full supply during shortages, while junior appropriators receive no water or only a partial supply. The correlative rights doctrine has been criticized as providing all users with an inadequate supply during shortages. However, the egalitarian sharing feature of correlative rights has received considerable scholarly admiration. Indeed, the Restatement water conflict criteria may be interpreted as an effort to supplant the correlative rights doctrine with a more formal system of criteria to use in making judicial water allocation decisions in times of shortage.

Federal Equitable Apportionment

Equitable apportionment is the doctrine used by U.S. Supreme Court to settle interstate water conflicts. Equitable apportionment is also a doctrine of international law used to resolve international water conflicts. In either context, the court uses the equable doctrine to resolve water conflicts between what essentially are two sovereign legal entities. As developed by U.S. Supreme Court, equitable apportionment is fact intensive, meaning that what is equable in a particular case depends very strongly upon what the facts of the case are. However, in conflicts between Western states, the court has followed a modified version of prior appropriation. Significantly, in a 1982 decision, the Court suggested that senior water allocated under a prior decree might be subject to reallocation to a more beneficial or more efficient junior use (*Colorado v. New Mexico,* 459 US 176, 1982). In a 1984 decision, the Court clarified that the state claiming that its new use would be more beneficial would need to prove that case by clear and convincing evidence, but otherwise left open the possibility that water could be allocated away from senior uses to more beneficial junior uses (467 US 310, 1984). The Court has also recognized the futile call doctrine that allows junior appropriators to use water out of priority when the water would not reach downstream senior appropriators in a timely fashion in usable quantities (*Nebraska v. Wyoming,* 325 US 589, 1945). The equitable apportionment doctrine has been characterized as prior appropriation with its rough edges smoothed off (Tarlock, 1985). However, the *Colorado v. New Mexico* feature of judicially reallocating water from senior uses to more beneficial junior uses is clearly a dramatic (and overdue) departure from the priority doctrine of prior appropriation.

RESTATEMENT OF TORTS WATER CONFLICT RESOLUTION PRINCIPLES

The Restatement of Torts reflects the most systematic attempt to identify relevant equable factors to be judicially considered to resolve water-use conflicts. In this section the Restatement principles are developed, and their potential relevance to resolve the new water use conflicts involving dormant rights, tributary groundwater, and environmental disputes are explored.

Purpose of the Interfering Use

In traditional water-use conflicts, the new use is the interfering use. Consequently, this factor may be interpreted as favoring the status quo, including the senior water use. Traditional western water conflicts are often between irrigators, so the purpose of the new interfering use would be considered on an equal plane with the senior irrigation uses. This could also hold true for most dormant claims and tributary groundwater conflicts. However, if the "new use" is an environmental use, it can be argued that the environmental use is the temporally senior use, even if the environmental use is not legally accorded a senior appropriation. In most cases the environmental or habitat uses would have occurred prior to settlement and therefore could claim a natural senior temporal priority.

Suitability of the Interfering Use to the Watercourse

This factor addresses the situation in which the new use would take all or most of the water from the stream. Under the riparian notion that all riparian landowners have an equal right to use the stream (which does not mean they get an equal share of the water, however), it would be unfair for one user to monopolize the available supply. This factor could be used in reverse in which a senior user has monopolized the stream, suggesting that supplies should be reallocated to obtain a more equitable distribution of available supplies. This factor could also be turned against senior appropriators in environmental disputes, in which the argument is that the senior appropriators are making a disproportionate demand on the stream, given the environmental disruption occasioned by the senior diversions.

Economic Value of the Interfering Use

One of the factors courts consider in addressing a variety of natural resource disputes is the economic impact. Does the new interfering use have substantial positive economic impacts that compensate for the harm to other riparians? Conversely, does taking water away from senior appropriators for environmental purposes have unacceptable economic consequences?

Social Value of the Interfering Use

This is a direct equitable consideration. What are the competing social policies or values supporting one use or the other? If the uses are all irrigation, the social value may be the same. If the new use is endangered species protection, does that social imperative deserve special consideration? If the environmental use is considered to have a natural temporal seniority as discussed earlier, does the irrigation or other diversion have a social value that justifies its interference with the environmental use?

Extent and Amount of Harm Caused

Many riparian cases deal with the issue of if the new use diminishes the natural flow of the stream to the detriment of other riparians who do not use any water but who derive pleasure from knowing that the stream is flowing undiminished through their land. Typically, the nonusing riparians receive little consideration in these circumstances. Other cases deal with the more difficult issue of mutually exclusive uses: if I use the water you therefore don't receive enough to continue your use. It is clearly relevant to the judge if the interference is minimal, significant, or catastrophic.

Practicality of Adjusting the Water Uses

If both competing water uses can be accommodated through a physical solution as discussed earlier, or by some other adjustment of water uses, the policy would be to accommodate as many uses as possible. If the amount of interference is small, there is a significant opportunity to accommodate all uses through adjusting both uses. If the uses are truly mutually exclusive, the problem is more difficult. However, courts are always interested in finding a win-win solution if possible, which is the purpose of this criterion.

One wonders if one way of adjusting competing water uses might be to require both users to achieve a higher level of use efficiency. If the senior use is old and relatively inefficient, is it not reasonable to consider requiring the senior user to upgrade the method of use to current standards? Perhaps, as the final criterion suggests, the senior user should not be required to upgrade the water-use efficiency if doing so is beyond the senior's economic reach (or if the new users can afford to do so out of their profits from their new water use). But in many circumstances the senior's use, which may have been state of the art when originally undertaken, is now far behind the times. In environmental policy, polluters are generally given only a limited amount of time to bring their pollution control facilities up to current technological standards. There seems no good reason from exempting water users from this same expectation.

Protecting Existing Values

In the narrowest sense, this factor considers protecting existing property or economic values and probably favors maintaining the status quo. However, if the notion of value is broadened to include noneconomic values, such as social or environmental values, the outcome may be different. Then one is left with managing the trade-off between protecting existing economic values versus protecting existing social or environmental values.

Justice of Requiring the User Causing the Harm to Bear the Loss

An important issue in western groundwater law is the issue of protecting the means of diversion. In a famous Colorado case, a municipality was required to pay an irrigator to deepen his senior well when the junior municipal wells lowered groundwater levels. The theory of the case was, in simple terms, that the municipality had substantial financial reserves and could afford to provide the senior appropriator a new well, and consequently that it would be unfair not to require the municipality to do so. The alternative would have been to deny the municipality its junior wells because they were interfering with senior wells (*Colorado Springs v. Bender*, 366 P2d 552, 961). If, however, the new use were not profitable enough to enable the junior user to easily afford compensating the senior users, the basis for resolving the conflict is not so clear-cut.

One imagines that judges operating under the Restatement would prefer to see if there is some way to accommodate both uses, through adjusting both uses (i.e., giving each user less water) or through a physical solution (i.e., engineering) that provides a fair share to each party. The new user may be required to pay for the physical solution, or perhaps the parties would share the costs, again depending upon the circumstances. If this outcome is not possible, the Restatement leaves the door open for the judge to allocate water away from the senior user to the new or "interfering" user if the new use is suitable to the watercourse, has a greater social value, economic value, or both than the senior use, or results in an overall more beneficial use of the water. However, as in *Colorado v. New Mexico* (459 US 176, 1982), discussed at pages 62–3, the new user has to make its case strongly in order to have water reallocated in its favor. The Restatement has several factors that inherently favor maintaining the status quo, reflecting the law's naturally conservative bent. The benefits of change need to be substantial in order to justify departing from the status quo, especially if the change is uncompensated. However, when that condition is met, the status quo is not an absolute barrier to change. That is the salient policy point.

COLORADO CONJUNCTIVE-USE REQUIREMENTS

As noted earlier, the state of Colorado has made the most progress in integrating the competing uses of surface and groundwater (MacDonnell, 1988). Indeed, Colorado has wrestled most vigorously in attempting to reconcile the protection of vested senior surface appropriations with the maximum utilization of groundwater by junior appropriators. The Colorado conjunctive-use requirements approach basically reflects the farthest that any Western state has gone in an attempt to reconcile senior surface appropriations and competing junior appropriations to tributary groundwater. The Colorado approach does not depart from traditional appropriation principles: instead, it aggressively pushed the concept of the physical solution in order to accommodate competing surface and groundwater uses. As such, it is an important precedent, as most water policymakers would be inclined to emulate the Colorado approach before attempting more radical policy options.

Colorado basically follows a policy of giving junior groundwater appropriators several methods for providing water to the stream (or directly to senior surface appropriators) when junior wells reduce streamflows. Under conventional priority administration, the junior groundwater appropriators could simply be issued closing orders and stopped from pumping until the senior surface appropriators uses are satisfied. Colorado provides numerous options for junior groundwater appropriators to avoid the threat of being issued a closing order altogether. Junior groundwater appropriators may transfer senior surface priority dates to their junior wells as a way of legally protecting those wells. Junior groundwater appropriators are also entitled to meet priority calls by providing substitute water to senior surface appropriators. Finally, under the futile call doctrine, junior wells are not subject to priority administration if the resulting increase in streamflow would not occur until after the senior's (typically) irrigation water need has passed.

Nonetheless, one can legitimately argue that even the legal accommodation of junior Colorado groundwater appropriators, which go much further than any found in any other Western state, places too great a burden on junior groundwater appropriators and too small a burden on senior surface appropriators (MacDonnell, 1988). The plan of augmentation can impose considerable expenses on junior groundwater appropriators, with no corresponding obligation on senior surface appropriators. One can argue that senior appropriators should be required to make some sacrifice of their own: they should be required to make water-use efficiency improvements as a condition for being able to call out junior wells (MacDonnell, 1988).

How likely is Colorado to impose use efficiency requirements upon surface appropriators? Currently a combination of factors has made it easier for junior groundwater appropriators to satisfy senior calls (MacDonnell, 1988). If circumstances change and junior groundwater flow augmentation requirements increase, Colorado water policymakers might be forced to address the efficiency issue. This policy evaluation could be triggered by Colorado's loss to Kansas on the Arkansas River compact litigation (115 Sup. Ct. 1733, 1995).

It is unlikely that many Colorado surface irrigators, with nineteenth-century priority dates, are using twenty-first-century irrigation practices; suggesting that junior groundwater appropriators might need to supply less water to senior surface appropriators if senior surface appropriators needed less water to begin with. As a minimum, a review of surface irrigation practices to reduce gross waste of water (e.g., flood irrigation vs. furrow, gated pipe, or sprinkler irrigation) would certainly be in order. Flood irrigators (who irrigate by flooding the entire field at once) could be required to adopt more efficient irrigation water distribution systems.

PLATTE RIVER COOPERATIVE AGREEMENT

The Platte River Cooperative Agreement is the result of decades of conflict over endangered species and water diversions from the Platte River. Irrigation and other water impoundments and diversions in Colorado, Wyoming, and Nebraska have depleted Platte River streamflows for endangered species in Nebraska. With federal officials poised to withhold approval of Platte River water-use permits in Nebraska, Colorado, and Wyoming pending resolution of Platte River endangered species issues, the parties reached a compromise (Aiken, 1999). The states agreed to provide water, land, and money for endangered species protection and to subject all new water uses to endangered species requirements. The federal government agreed to temporarily lower its endangered species water requirements, to authorize the continuation of existing water uses that were subject to federal approval, and to provide one-half the funding for endangered species protection.

The endangered species water requirement has been a difficult issue to resolve. Federal officials have claimed that 417,000 acre-feet per year were needed for the successful recovery of Platte River endangered species (Aiken, 1999). The states disputed that figure as being far too high. As a compromise the parties agreed to take an incremental approach, establishing a goal of providing 130,000 to 150,000 acre-feet of new water (roughly one-third) for endangered species during the first decade of the cooperative agreement's operation. If experience indicates that this was sufficient for successful recovery of Platte River endangered species, no additional water would be required. If more water were needed, however, a new agreement would have to be negotiated (Aiken, 1999). The additional water will come primarily from voluntarily improving the efficiency irrigation water use, and from purchasing water rights from Platte surface water irrigators. Nebraska irrigation districts are providing 70,000 acre-feet per year and are depositing the water into an "environmental" water storage account. The irrigation districts will generate their water deposits into the environmental account by improving irrigation practices within the districts and by reducing district water transmission losses. At this point, there will be no mandated irrigation water-use efficiency requirements.

Will the Cooperative Agreement succeed? That remains to be seen. Wyoming and Colorado must implement costly water supply projects to provide their share of augmentation water. In Nebraska, it is not yet legal to purchase irrigation water rights and use the water for endangered species protection. If these significant obstacles can be overcome, and the first increment of 130,000 to 150,000 acre-feet of new water for endangered species is successfully acquired, the issue will be if the additional water will allow sufficient endangered species recovery for federal officials to feel comfortable not to request additional water for endangered species protection. Another factor is whether potential Congressional modifications to the federal Endangered Species Act relax endangered species protection requirements sufficiently to render the Cooperative Agreement moot.

CONCLUSION

Priority of appropriation is the cornerstone of the appropriation system of water rights: when insufficient water is available, senior appropriators are entitled to their complete allotment, and junior appropriators (i.e., newer users) are entitled to what is left, if anything. There is no requirement for the senior appropriator to share with juniors during shortages, and generally no requirement to meet any more than nominal water-use efficiency requirements, even during shortages. This system clearly favors the early senior appropriators, but does nothing to balance junior needs and senior rights.

The priority doctrine is still the cornerstone of western water law. The advent of water marketing has provided a relatively convenient and effective way for junior appropriators to acquire needed water supplies: buy senior water rights and convert the water use to the junior use. However, water marketing may not be enough to resolve new western water disputes that result from a fundamental miscalculation and subsequent misallocation of water needs, uses and rights. Shocks to the water allocation include 1) the exercise of dormant Indian or federal rights, 2) stream depletion from junior tributary wells, and 3) environmental needs.

The Colorado system for dealing with tributary groundwater disputes and the Platte River Cooperative Agreement are two recent approaches for dealing with these water system shocks. Under the Colorado tributary groundwater regime, Colorado has some technical water law changes that facilitate dealing with disputes between surface water users and tributary groundwater users. The most prominent is allowing an appropriator with both a surface appropriation and a tributary groundwater appropriation to transfer the senior surface-water priority date to the junior tributary well. This allows the junior well to become a senior well and gives the senior surface appropriator a more secure irrigation water supply. However, the two main features of the Colorado system for dealing with surface-tributary groundwater disputes are augmentation requirements and water marketing. To avoid a priority call, junior tributary well owners can provide sufficient replacement water to the state engineer to meet the needs of senior surface appropriators who would otherwise be able to call out the junior wells (and force their use to be discontinued). Flow augmentation requirements can be met by impoundment, by pumping groundwater directly into a senior surface appropriator's diversion canal, and by water marketing. The latter involves purchasing and in effect retiring surface appropriations. The senior use that was purchased can no longer make a priority call, and the surface water can be used to supply the needs of other senior surface appropriators.

The Colorado approach represents an imaginative use of existing water management tools, notably water marketing. The Colorado approach does not, however, use the option of requiring irrigators (senior and junior, surface and tributary alike) to improve the efficiency of their use as one way of dealing with an overappropriated supply. If a senior surface appropriator is allocated less water because of higher state irrigation efficiency standards, the amount of water that is needed from junior tributary appropriators to satisfy senior rights is reduced. Similarly, if junior tributary appropriators are allocated less water because of higher state irrigation efficiency standards, the extent of tributary junior appropriator water use and the associated interference with senior surface appropriators is also reduced. Although the magnitude of these savings may be marginal at the individual irrigator level, collectively it may be significant.

Colorado may be forced to consider the irrigation efficiency issue as a way to deal with losing the Arkansas River litigation to Kansas. As an intermediate approach, Colorado could offer irrigators financial incentives to improve irrigation practices to reduce overall irrigation water demands. The incentives approach could be a step toward imposing mandatory state irrigation efficiency practices, as Arizona has done to cope with groundwater depletion.

The Platte River Cooperative Agreement takes a similar approach to deal with increasing water supplies for endangered species. Colorado, Wyoming, and Nebraska will provide new water for endangered species by developing surface and groundwater irrigation efficiency improvements. The states will also provide financial incentives for water conservation practices that yield new water for endangered species. If water marketing is legally authorized in Nebraska, purchasing surface irrigation appropriations and using the water for endangered species will also be possible.

The Platte River Cooperative Agreement does address the improved water-use efficiency issue to a larger degree than the Colorado tributary groundwater program. However, there is another difference between the two water management scenarios that deserves additional discussion. In Colorado the dispute is largely between irrigators using surface water and irrigators using tributary groundwater. In the Platte River scenario, the main dispute is between irrigators and endangered species. In the Colorado tributary groundwater scenario, a correlative rights solution may be fair under which all users must become more efficient in order to accommodate all uses to an inadequate supply. In the Platte River case, however, the public has decided through public policy (the Endangered Species Act) that water use for endangered species is more important than any other water use. It may be reasonable under these circumstances to expect irrigators to make some reasonable accommodations for endangered species up to a point, for example, by improving irrigation efficiency practices up to a reasonably current level. Beyond that point, however it seems unfair to impose the entire cost of the public's preference for endangered species protection upon irrigators, and irrigators should reasonably expect the public to pay for any additional water for endangered species, either directly through water marketing or indirectly by paying for further improvements in irrigation water-use efficiency. Although the public has accepted some businesses going out of business rather than bear the expense of installing required pollution control equipment, it seems unfair to put irrigators in that same category.

A first step in establishing improved irrigation water-use efficiency requirements would be to review existing surface water uses and require very wasteful uses to improve their efficiency practices, similar to the Imperial Irrigation District case. This might be sufficient in many cases to solve any water supply shortages. If not, the next step would be to require senior surface appropriators to upgrade irrigation practices to current irrigation waste distribution systems and management practices, as in the Washington case. This would be more costly and more controversial, but could yield needed water for higher-priority uses in some cases. Improved surface irrigation efficiency could first be encouraged through financial incentives, such as cost-sharing assistance or low-interest loans. The incentives could be coupled with ultimately mandatory requirements, similar to the Arizona groundwater depletion program (in which irrigators have ten years to comply with irrigation efficiency requirements).

The standard argument to this approach is that requiring surface irrigators to be more efficient may modify traditional return flow patterns, which is a fair statement. However, in most cases all surface appropriators are subjected to a wide variety of water-related uncertainties. The irrigator's crop choices and planting date determine what the crop water needs are during the growing season. Some crops require more water than others; some crops can wait for water longer than others. The timing and amount of precipitation is a major factor: it influences if the farmer needs to irrigate, how much the farmer needs to irrigate, and how much water is in the stream available for diversion. Precipitation also determines how much water stored in reservoirs is available in a particular growing season. All these weather and crop production variables also apply to all irrigators upstream and downstream. This further complicates the situation by influencing when upstream irrigators will irrigate, when downstream appropriators will irrigate, and when downstream irrigators make priority calls.

Amid all this uncertainty, it makes little sense to contend that the possible change in return flow from improved upstream irrigation efficiency is a crucial element that cannot be modified under any circumstances. Indeed, if efficiency requirements are imposed on all surface irrigators, changes in return flow patterns may be offset by other benefits from increased water-use efficiency. If upstream appropriators are entitled to less water due to efficiency requirements, downstream appropriators are similarly entitled to less. Thus while there may be less return flow coming to me, I may need to pass on less water to downstream senior appropriators. In addition, when upstream appropriators divert less water, more water is passed down to me.

Finally, if a state water policymaker determines that certain irrigation practices are wasteful, remember that no person acquires any water rights in another person's wasteful use.

There is little justification for allowing inefficient water uses to continue unregulated, when there are important unmet water needs. This is tantamount to legislating that existing polluters need not control their pollution but all new polluters must strictly control their pollution. The existing polluters are the ones who caused the pollution problem, not future polluters, and the problem will not be solved until all polluters control their emissions. Similarly, water shortages are caused by existing users, all of whom should be required to play a role in redressing the imbalance between water supply and water needs by using less through more efficient practices.

The continuing competition for water in the Great Plains and the West demands that all users be held to some standard of water-use efficiency. Water marketing can be a fairly easy policy fix to accommodate new uses through buying water rights from existing users. However, the availability of water marketing should not blind policymakers to the need for establishing realistic water-use efficiency standards for all users in order to establish a more equitable distribution of our scarce and valuable water resources.

REFERENCES

Aiken, J. D. 1980a. The National Water Policy Review and Western Water Rights Reform: An Overview. *Nebraska Law Review* 59, no. 2: 327–44.

Aiken, J. D. 1980b. Nebraska Ground Water Law and Administration. *Great Plains Natural Resources Journal* 59, no. 4: 917–1000.

Aiken, J. D. 1999. Balancing Endangered Species Protection and Irrigation Water Rights: The Platte River Cooperative Agreement. *Great Plains Natural Resources Journal* 3, no. 2: 120–58.

American Law Institute (ALI). 1979. Interference with the Use of Water, III. *Restatement (Second) of the Law of Torts.* Philadelphia: American Law Institute, 180–270.

Dunning, H. C. 1986. The "Physical Solution" in Western Water Law. *University of Colorado Law Review,* 56, no. 3: 445–83.

Dunning, H. C. 1987. State Equitable Apportionment of Western Water Resources. *Nebraska Law Review* 66, no. 1: 6–119.

MacDonnell, L. J. 1988. Colorado's Law of "Underground Water": A Look at the South Platte Basin and Beyond. *University of Colorado Law Review* 59, no. 3: 579–624.

Pring, G. W., and K. A. Tomb. 1979. License to Waste: Legal Barriers to Conservation and Efficient Use of Water in the West. *25 Rocky Mtn. Min. L. Inst.* New York: Matthew Bender.

Reisner, Marc. 1986. *Cadillac Desert: the American West and its Disappearing Water.* New York: Penguin Books.
Reisner, Marc, and Sara Bates. 1990. *Overtapped Oasis: Reform or Revolution for Western Water.* Washington DC: Island Press.
Tarlock, A. D. 1985. The Law of Equitable Apportionment Revisited, Updated, and Restated. *University of Colorado Law Review* 56, no. 3: 381–411.
Tarlock, A. D. 1999. *Law of Water Rights and Resources.* St. Paul, Minn.: Clark Boardman.

Water Across Borders: Judicial Realities

Peter J. Longo

BORDER REALITIES

Kansans have been known to tell Nebraskans that the best things about Nebraska are the signs reading "You are now leaving Nebraska." Likewise, Nebraskans have been known to utter the converse to Kansans. Undoubtedly, but for the boundary signs, it is difficult to notice geographical differences from state to state to province to province over the vast Great Plains outlay. It truly seems that the telephone pole is the native tree and boundaries only exist because of a highway marker. As many an impatient traveler of the Great Plains has thought or said, "so much looks the same." One thing that is certain, however, is that in the political arena, borders are clearly recognized. Political scientists have devoted serious attention to borders and the human tragedies resulting from border disputes (Cashman, 1993). Although loss of life and suffering from the violence of war seem unlikely, conflict over water does have a price and the search for water heightens the border recognition.

Borders have long presented interesting possibilities for the study of both unity and conflict. If the citizens of the Great Plains shared an overriding common political culture or economic drive, then there would be no apparent need for political boundaries. However, such is not the reality, and boundary disputes persist not only on the Great Plains, but also in many other regions in the United States and beyond (Civic, 1998; Davis, 1999). Greater and greater stress is being placed on the most precious natural resource, water. The citizens and lawmakers of the Great Plains increasingly scrutinize water use or nonuse. Water conflicts are a drain on economic as well as natural resources and the conflicts tend to bring the worse out of parochial actors. The fights between Nebraska and Kansas are ongoing, as a

June 29, 2000 Supreme Court opinion illustrates (*Kansas v. Nebraska and Colorado,* 120 Sup.Ct. 2764, 2000) and this prolonged dispute has captured the attention of non-Plains observers. For example, the *Atlanta Constitution* provided the following:

> Nebraska, Colorado and Kansas signed a deal in 1943 that spelled out how much water each state could use from the Republican River. But it's been anything but amicable among the three states. Saying Nebraska wasn't abiding by the compact, Kansas sued its northern neighbor in May. Kansas claims that Nebraska has been siphoning off too much water from the river for irrigation, leaving it some 10 million gallons short every year. The squabble may be an omen for Georgia, Alabama, and Florida, which are in intensive negotiations to divide the rivers they share . . . but the conflict between Nebraska and Kansas shows that such agreements don't necessarily settle water wars. (Seabrook, 1998, 6E)

In time of conflict, it is unlikely that a state legislator would argue for giving a neighboring state more water than her local constituents. Thus by default, depositories of border conflicts over water are the courts. If not for the guidance of the courts, water issues would be a divisive force in and between state governments on the Great Plains, as these local entities are often viewed as the protectors of their own particular citizens. Of course, if this is taken to the logical extreme, then tragedy can result as a state collects water without any concern for bordering neighbors. Such potentially tragic environmental consequences can occur akin to the situation described by Hardin (1968) in his classic, "The Tragedy of the Commons." In the following passage, draw the analogy to citizens removing water from the Great Plains:

> Picture a pasture open to all. It is to be expected that each herdsman will try to keep as many cattle as possible on the commons . . . As a rational being, each herdsman, seeks to maximize his gain . . . Adding together the component partial utilities, the rational herdsman concludes that the only sensible course for him to pursue is to add another animal to his herd. And another, and another . . . But this is a conclusion reached by each and every herdsman sharing a commons. Therein is the tragedy. Each man is locked into a system that compels him to increase his herd without limit—in a world that is limited. (1243)

The situation described by Hardin has real and meaningful messages for Great Plains jurisdictions. It is important to know how political institutions such as the courts monitor the water on the Great Plains as the situation is clearly analogous to the commons. With respect to allocation, neighbors are encouraged to take their fair share only because of judicial pronouncements. As we will see in this chapter, neighboring states are often pitted against each other. It is truly the individual state courts that demand for their citizens their share of water. "Nobody wants to sue a neighbor, but sometimes the neighbor gives us no choice," says Kansas Attorney General Carla Stovall (Anderson, 1998, 1). Attorney General Stovall is clearly correct, and this chapter, case after case, tediously illustrates neighborly conflict and the role of the judiciary in settling conflict.

ILLUSTRATIVE CASES

Kansas versus Colorado

Conflict is at the heart of neighborly reaction when the story involves water. Such is the story of a 1906 case, *Kansas v. Colorado* (206 U.S. 46). Kansas viewed Colorado's use of the Arkansas River as threatening the flow of the river through Kansas. Naturally, Colorado viewed the Arkansas River as belonging to the citizens of Colorado. Despite the obvious nature of the conflict, this case serves as a poignant reminder of the complicated nature of water cases. Indeed, the majority opinion cautions that

> [t]he testimony in this case is voluminous, amounting to 8,559 typewritten pages, with 122 exhibits, and it would be impossible to make a full statement of facts without an extravagant extension of this opinion, which is already too long. (106)

For the student of water law it is important to read this caution with a bit of wonder. This wonder ought to lead to an appreciation of the complexity of water cases. Imagine typing, without the use of a word processor, 8,599 pages. The mechanical nature of typing the case foreshadows the legal complexity involved in this boundary dispute. I will limit the analysis to the central issue. The central issue as offered by the Court is as follows:

> Turning now to the controversy as here presented, it is whether Kansas has a right to the continuous flow of the waters of the Arkansas River, as the flow existed before any human interference therewith, or Colorado the right to appropriate waters of that stream so as to prevent that continuous flow is subject to the supervisory authority and control of the United States. (185)

Simplified, the Court queries whether or not Colorado and Kansas can share water and if the Court can serve as the final arbitrator of this conflict. With regard to the latter, the Court clearly serves as the final arbitrator, and to the former issue the Court holds: "One cardinal rule, underlying all the relations of the states to each other, is that of equality of right. Each state stands on the same level as all the rest" (97). In water battles between the Great Plains neighbors (particularly states) the U.S. Supreme Court urges the neighbors to be neighborly. The Court touts the water use of Colorado, finding "that the result of that appropriation has been the reclamation of large areas in Colorado, transforming thousands of acres into fertile fields" (117). Further, the Court agrees that the use diminished the availability of water for Kansas, but it "has worked little, if any, detriment . . . we are not satisfied that Kansas has made out a case entitling it to a decree" (117). The Court again is in support of water use, but only in a fair manner. To placate Kansas's concerns and to assure judicial fairness, the Court concluded:

> [If] the depletion of waters of the river by Colorado continues to increase there will come a time when Kansas may justly say there is no longer an equitable division of benefits and may rightfully call for relief against the action of Colorado, its corporations and citizens, in appropriating the waters of the Arkansas for irrigation purposes. (117)

Thus, this early case sends the message that the Great Plains citizens ought to be encouraged to use water in an efficient and fair manner. Clearly, this case suggests that the Supreme Court is willing to determine neighborly behavior on an ongoing basis.

Nebraska versus Wyoming

Just as in the examples found in previous chapters and the preceding case, there seems to be a desire for competing interests to hoard as much water as possible, regardless of the impact upon neighbors. Such has been the scenario in water disputes between Nebraska and

Wyoming. The two states argued before the Supreme Court in 1945 and 1995. The 1945 case of *Nebraska v. Wyoming* (325 U.S. 589) further exemplifies the tensions that exist between states. In this case, Nebraska argued that Wyoming and Colorado were using more than their fair share of water and thus did not allow enough water to enter into the flow of the heritage-rich Platte River.

Justice Douglas wrote the opinion for the majority and as we will see, his opinion offered a careful approach of balance for the affected states. Of interest to our study is how Douglas so aptly captured the essence of this vital Great Plains river, the Platte. He wrote:

> The North Platte River rises in Northern Colorado in the mountainous region known as North Park. It proceeds in a northerly direction on the east side of the Continental Divide, enters Wyoming west of Cheyenne, and continues in a northerly direction to the vicinity of Casper. There it turns east across the Great Plains and proceeds easterly and southerly into and across Nebraska. About 40 miles west of the Nebraska line it is joined by the Laramie River. At North Platte, Nebraska, it is joined by the South Platte, forming the Platte River. It empties into the Missouri River at Plattsmouth, near the western border of Iowa. In eastern Wyoming it gradually broadens out, losing velocity. In western and central Nebraska its channel ranges from 3000 to 6000 feet; it frequently divides into small channels; and in times of low water is lost in deep sands of its bed. Here it is sometimes characterized as a river "two miles wide and one inch deep." (593–94)

It is probably difficult for those who have not experienced the paucity of water on the Great Plains to imagine why one would actually fight over a river "two miles wide and one inch deep." Yet, the scarcity of water is a constant reminder of the struggle for survival that nearly all residents have encountered. For an obvious outsider, Justice Douglas displays a keen sensitivity to the demands made by the Plains litigants for this most precious resource.

Wyoming and Colorado claimed that Nebraska did not present a "justiciable controversy" or, in other words, the conflict did not portray any determinable injury. Justice Douglas disagreed and stated:

> What we have then is a situation where three states assert against a river, whose dependable natural flow during the irrigation season has long been over appropriated, claims based not only on present uses but on projected uses as well. The various statistics with which the record abounds are inconclusive in showing the existence or extent of actual damage to Nebraska. But

we do know that deprivation of water in arid or semi-arid regions cannot help but be injurious. (610)

Thus, Justice Douglas and the majority of the Supreme Court saw the justification for constant supervision of water allocation. This supervision was taken from the directions of a Special Master— generally a neutral actor who provides the technical expertise to the various litigants. It is the Special Master's task to balance opposing claims within the realities determined by hydrology. The courts either endorse the Special Master's finding or determine that the Special Master did not follow the fairness directive of the court involved. In this case, the majority said, "We conclude that the flat percentage method recommended by the Special Master is the most equitable method of apportionment" (646). The dissent took exception to the majority's willingness to serve as an umpire for water issues. Justice Roberts's dissent said:

> The future will demonstrate in my judgment, how wrong it is for this court to attempt to become a continuing umpire or a standing Master to whom the parties must go at intervals for leave to do what, in their sovereign right, they should be able to do without let or hindrance, provided only that they work no substantial damage to their neighbors. In such controversies the judicial power should be exercised upon proper occasion, but as firmly withheld unless the circumstances as plainly demanded by intervention of the court. Such mutual accommodations for the future of Nebraska and Wyoming should be arranged by interstate compact, not litigation. (658)

Justice Roberts's restrained position places unfounded faith in other political actors. As we will soon see, this faith was baseless, as the dispute was brought to the Supreme Court in 1995.

Nebraska versus Wyoming II

In *Nebraska v. Wyoming* (515 U.S. 1, 1995) the U.S. Supreme Court was once again called upon to settle the claims between Nebraska and Wyoming. Like the 1945 case, the case at bar presented some complicated facts and issues. Once again, the ongoing concern regarding the factors impacting the diminishing Platte River flows served as the core of the case. A new twist to this case concerns environmental issues rather than purely irrigation demands. The environmental claim was sent to the Special Master with the following language:

> If Nebraska is to have a fair opportunity to present its case for our doing so, we do not understand how we can preclude it from setting forth that evidence of environmental injury, or consign it to producing that evidence in some other forum, since this is the only court in which Nebraska can challenge Wyoming's projects. And as for Wyoming's argument that any proof of environmental injury that Nebraska will present will be highly speculative, the point is urged prematurely. Purely speculative harms will not, of course, carry Nebraska's burden of showing substantial injury, but at this stage we certainly have no basis for judging Nebraska's proof, and no justification for denying Nebraska the chance to prove what it can. (12, 13)

Nebraska was called upon to build its case. Underlining this call is the Court's concern for fair allocation of water between states. In 2001, Nebraska and Wyoming officials did reach agreement for fair allocation. Further legal disputes and resulting costs will apparently cease. The court's push for fairness can bring state officials together. Other cases further highlight the court's concern for fairness.

Texas versus New Mexico

Texas v. New Mexico (462 U.S. 554, 1983) illustrates the point that water compacts, despite the inherent effort of interstate cooperation, are often matters for judicial interpretation. The 1948 Pecos Basin Compact was intended to fairly provide water from the Pecos River to the citizens of Texas and New Mexico. The Pecos River is the principal river in Eastern New Mexico and flows southerly until joining the Rio Grande in Langtry, Texas (556). Like nearly every source of water, the Pecos is highly coveted by many suitors. This attention exists despite the rather inadequate nature of the river. Justice Brennan, writing for the majority, described the river as follows:

> The Pecos barely supports a level of development reached in the first third of this century. If development in New Mexico were not restricted, especially the groundwater near Roswell, no water at all may reach Texas in many years. As things stand, the amount of water Texas receives in any year varies with a number of factors besides beneficial consumption in New Mexico. These factors include, primarily, precipitation in the Pecos Basin over the preceding several years, evaporation in the McMillan and Alamogordo Reservoirs, and

the non-beneficial consumption of water by salt cedars and other riverbed vegetation. (557)

It is the harsh reality of this parched Plain that arouses the interest of suitors. Quite understandably, states are most unwilling to relinquish water to neighbors given the paucity of supply. However, compacts do attempt to fairly allocate the resource.

However, as this case reminds us, not all issues are settled by the Compact. Judicial intervention forces fairness. On this point the Court stated:

> Texas' right to invoke the jurisdiction of this Court was an important part of the context in which the Compact was framed; indeed, the threat of such litigation undoubtedly contributed to New Mexico's willingness to enter into the compact. (569)

In keeping with the spirit of cooperation, the Court ordered the Commission to determine the amount of water Texas needed and if shortfalls were due to "man's" activities (575). In other words, the Court mandated cooperation and fairness and the Compact failed to deliver such a result.

Oklahoma and Texas versus New Mexico

A 1991 case involving three Plains States can be found in *Oklahoma and Texas v. New Mexico* (501 U.S. 221). Once again, this case illustrates that even when Plains States enter into compacts to share water, the interpretation of such contracts often requires judicial resolution. Oklahoma and Texas brought this action against New Mexico because of the latter state's storage of water not specifically mentioned by the original compact. Like so many water cases, this dispute includes an incredible factual maze; I provide only the relevant facts. The river in dispute is the Canadian River, and for the interested onlooker of the Great Plains, the physical description provided by the pen of Justice White enhances the imagery of the Plains. Justice White wrote of the Canadian River:

> An interstate river which arises along the boundary between southeastern Colorado and northeastern New Mexico, in the vicinity of Raton, New Mexico. From its headwaters, the Canadian River flows south to the Conchas Dam in New Mexico, then generally east for 65 river miles to the Ute Reser-

voir in New Mexico, and then into the Texas Panhandle. After traversing the panhandle, the river flows into Oklahoma where it eventually empties into the Arkansas River, a tributary of the Mississippi. (224)

The facts provide imagery of a life-giving river trudging through the hot and dusty High Plains with thirsty citizens trying to squeeze as much water as possible from the river before it seems to be nothing more than a mirage on the horizon. Such imagery was actually grounded in reality, as the three states signed a 1950 Canadian River Compact setting limitations on the amount of water that could be stored and used by each (225). This case was precipitated by New Mexico's unwillingness to count "spill waters" that were deposited in a de-silting pool. The de-silting pool was physically located in part of New Mexico's 200,000-acre-feet limitation. However, the 1950 compact did not provide specific guidelines for the de-silting pool.

New Mexico's position was that the issue of de-silting should be left to New Mexico, not the Compact Master, and certainly was not to be an issue for the Court. The majority of the Court disagreed with New Mexico's position and clearly held that it was the duty of the Court to settle controversies between the States "over the water of interstate streams" (241). More precisely, the Court emphatically concluded:

> There is no doubt that such a dispute exists in this case. Oklahoma and Texas have properly invoked this Court's jurisdiction, and there is no claim that the "de-silting pool" issue has been presented. Thus, we see no legal basis for the Master refusing to decide the question and instead sending it to the Commission. Thus, we remand the "de-silting pool" question to the Master for such proceedings as may be necessary and a recommendation on the merits. (241)

We are again left with the clear message that the Court is willing to settle water disputes. Further, the Court truly takes on the role of peacemaker, and as such, seeks to offer a settlement that encourages a fair and reasonable use of water.

Colorado versus New Mexico

A similar message with similar geographic references, but without the added feature of compacts, was provided in *Colorado v. New Mexico* (467 U.S. 310, 1984). Colorado sought an equitable apportionment of the Vermejo River, which originates in Colorado and flows into New Mexico. Historically, all of the river's waters have been used exclusively by farmers and industrial users in New Mexico (312). Colorado sought to use water from the river to support industrial purposes. A Special Master was appointed to determine if Colorado could provide sufficient conservation practices to ensure an adequate supply of water for downstream users. In this case, the Special Master failed to ascertain the necessary conservation precautions and, in turn, Colorado lost its claim to the water. Specifically, the Court held:

> We continue to believe that the flexible doctrine of equitable apportionment extends to a State's claim to divert previously appropriated water for future uses. But the State seeking such a diversion bears the burden of proving, by clear and convincing evidence, the existence of certain relevant factors. The complainant must show, for example, the extent to which reasonable conservation measures can adequately compensate for the reduction in supply to the diversion, and the extent to which the benefits from the diversion will outweigh the harms to existing users ... The Special Master struggled, as best he could, to balance the evidentiary requirement against the inherent limitations of proving a beneficial use. However, we do not find enough evidence to sustain his findings. (323–4)

The Court, as in many other cases, served as the final determiner of fairness. The role of the Court in these boundary cases remains consistent. Peace is maintained by the judiciary as well as the opportunity for Plains States to receive their day in court. Further, it does appear that conservation is critical in assessing competing claims. However, as we will see in the case of *Nebraska v. Sporhase* (208 Neb. 703, 1981), conservation cannot be an excuse for the parochial hoarding of water.

Nebraska versus Sporhase

Nebraska v. Sporhase (208 Neb. 703, 1981) is one of the most important boundary water cases of the Great Plains. The facts of the case simply involved the use of water across the boundary of Nebraska and Colorado. In this case, a farmer had irrigated land that geographically straddled Nebraska and Colorado. The farmer wanted to irrigate the land in Colorado using a well in Nebraska. The Nebraska Department of Water Resources rejected a permit for the farmer because the application requested water for use in Colorado, and Colorado did not provide for reciprocal use. In support of the Water Resources Director's position, the Nebraska Supreme Court held:

> We note that conditioning a landowner's right to transfer groundwater either within or without Nebraska does not deprive him a property right, since under Nebraska common law, groundwater may not be transferred off the overlying Nebraska land at all unless the public, owners of the water, grant that right. (710)

Consistent with the trappings of parochialism, the Nebraska Court forbade the transfer of water across borders. Although this may seem unneighborly, the court's rationale was based upon notions of preservation, rather than parochialism. On appeal, the U.S. Supreme Court rejected the public trust position and the resulting message of preservation. The Supreme Court held that water was an article of commerce and that Nebraska could not interfere with the federal commerce clause. The Court claimed that

> [t]he agriculture markets supplied by irrigated farms are worldwide. They provide the archetypical example of commerce among the several States for which the Framers of our Constitution intended to authorize federal regulation. (923)

The Supreme Court, the national judiciary, chose to reduce water to an article of commerce such that our Framers had somehow envisioned irrigated farms. This position raises serious obstacles to water conservation on behalf of the Great Plains regimes. Even when it appears that a Plains State is trying to conserve and protect her citizenry from environmental harm, there is no guarantee that this will be allowed solely in the name of environmentalism.

Arkansas versus Oklahoma

Indeed, a case involving a non-Plains State versus a Plains State, *Arkansas v. Oklahoma* (503 U.S. 91, 1992), provided a disappointing message for the citizens of the Great Plains. In this case, Arkansas was discharging waste into the Illinois River in accordance with an Environmental Protection Agency permit. The State of Oklahoma sought to prevent further discharge, as it violated Oklahoma water quality standards. Stating the obvious, the Court said, "Interstate waters have been a front of controversy since the founding of the Nation" (98). In this particular case, we find out that the State of Oklahoma, which enacted stricter environmental safeguards for the Illinois River, cannot place environmental safeguards which are more stringent than those placed by the EPA, and

> [t]hese provisions do not authorize the downstream state to veto the issuance of a permit for a new point source in another state, the Administrator retains authority to block the issuance of any state-issued permit that is "outside the guidelines and requirements of the Act." (102)

The loss for Oklahoma was an obvious setback for conservation and muddied the position that the Court was concerned about universal environmental and economic fairness. It is clear, from this excerpt, that water use is encouraged rather than discouraged. Given the scarce supply of water on the Great Plains, this case does not bode well for the citizens, present and future. We will see that Canada has a different view of conservation.

THE VIEW FROM CANADA

Interprovincial Co-ops versus The Queen

The message of conservation is more apparent in the Canadian Provinces than in the Plains States. A case that highlights this point is *Interprovincial Co-ops v. The Queen* (53 D.L.R. 3d., 321, 1975). In this case, the defendants operated chlor-alkali plants in Ontario and Saskatchewan. The pollutants caused fish kill in the waters of Manitoba. Naturally, Manitoba sought damages although the defendants claimed that it was beyond the scope of Manitoba to limit interprovincial acts. The facts of this case are remarkably similar to the facts of the Oklahoma-Arkansas dispute. The results are quite different. In the Canadian case, the court held:

> It is plain enough to me that a Province having rights in property therein is entitled to protect those rights against injury, and, similarly, to protect the interests that others may have in that property, by bringing or authorizing actions for damages, either at common law or under statutory provision. (335)

Thus, Manitoba can regulate water pollution because of the adverse effects upon a property interest, namely fish. This property interest is directly related to the environmental well-being of the citizens of Manitoba. Such regulation, unlike the Oklahoma case, does not disrupt federalism, as environmental protection is seen as a paramount concern for both levels of government. In the language of the court, the concern is not for mere punishment:

> Rather it is concerned with the protection and preservation of fisheries as a public resource, concerned to monitor or regulate undue or injurious exploitation, regardless of who the owner may be, and even in suppression of an owner's right of utilization. (335)

Lastly, the court clarifies the issue of which law ought to apply:

> Neither Saskatchewan nor Ontario can put forward as strong a claim to have their provincial law apply in the Manitoba action; in other words, the wrong in this case was committed or arose in Manitoba and not in Saskatchewan or in Ontario. (339)

But it would be a mistake to conclude that federal law is not utilized to ensure water equity. It is important to keep in mind that Canadian water law, as we will see throughout this chapter, tends to seek positive environmental results.

Canada versus Saskatchewan Water Corporation

In *Canada (Attorney-General) v. Saskatchewan Water Corporation* (103 D.L.R. 4th 250, 1993) federal law was used to support an injunction against two proposed dams in the province. The dams were to be constructed on the Souris River. As we have seen time and time again, neighbors, states, and provinces are reluctant to let any water pass beyond their own particular borders. With that as a reference point, the court's holding ought to be of little surprise. Precisely, the court held:

> The Souris River, which flows from Saskatchewan into North Dakota and then into Manitoba, is an international river. For this and other reasons, works associated with the river and its basin, fall within federal jurisdiction and are governed by federal law. (270)

This case clearly underscores the reality that provinces and states are not intrinsically well-suited to objectively deal with interborder water issues. Borders drive parochially based water policy and thus the court in this instance forced the borders to include more than one jurisdiction when deriving water solutions.

CONCLUSION

If the residents of the Great Plains were one big happy family, basking in shared culture and economic drive, there would be no future disputes and this chapter would be definitive of all Great Plains border water cases. However, the past dictates that this chapter is far from the last word on the subject. Clearly, the realities of case law suggest that more cases will percolate from the Great Plains. Perhaps it is too much to ask those who perceive an insatiable need for water to share water. Life is often harsh on the Great Plains, and without a continuous water supply, life simply would cease. Given the rather dramatic dynamic of perceived and real water needs, it is unlikely that legislators would take a position contrary to the electors. For example, one could only imagine the fate of an Oklahoma legislator telling constituents that she/he had just enacted legislation to give Texas more water from Oklahoma sources. Politics drives the lawmaker to hoard water or face the political backlash.

Fortunately, political human nature is balanced by judicial objective for face-saving water allocation. As the cases revealed, the courts provided an avenue for water-sharing deals to take form. The overriding message from the various courts is simple—water must be shared. This rather simple message ought to guide lawmakers as the citizens of the Great Plains become increasingly interdependent. Far too often, a common Great Plains heritage is claimed without legislation recognizing the common needs or appreciation. There exists some hope for neighborly cooperation. A short excerpt from the *Omaha World Herald* op-ed suggests that water wars are avoidable:

> Wyoming and Nebraska have reached a settlement in principle, at least, of their lengthy expensive dispute over the water in the North Platte River. . . .
> [I]f the deal is as workable as Governor Mike Johanns has indicated, perhaps it will set a precedent for Nebraska's water dispute with Kansas. . . . The settlement with Wyoming may or may not suggest the outline for an agreement

with Kansas. But all Nebraskans might well hope that it does, and that their elected officials try their hardest to work something out. What's at stake is the millions that would be spent on a lengthy fight, the sizable price tag of a trial and the possibility that Nebraska could lose and be required to pay money in reparations. (Op-ed, 2001, 26)

The accord between Nebraska and Wyoming was simply noted by a short *Rocky Mountain News* article on March 14, 2001, page 7A: "Nebraska and Wyoming agreed [March 13, 2001] to settle their $20 million battle over water in the North Platte River. The conflict dates back to the 1930s."

But when it comes to water disputes, it often seems that only the courts are able to draw the citizens together. Other lawmakers need to capture this cooperative spirit in the area of water law and policy.

Lastly, the comparative messages offered are insightful. The judicial pronouncements from the states do reflect an urging for cooperation. However, great deference is given to keeping water unencumbered, perhaps to the detriment of conservation. Canada seems more concerned with conservation. Given the great respect courts command, positive environmental messages would not only be well-received, but would ensure water resources for future generations. Such messages must be derived at other loci besides the courts.

REFERENCES

Anderson, Julie. 1998. *Omaha World Herald* (Omaha, Nebraska), May 27, 1.

Cashman, Greg. 1993. *What Causes War?: An Introduction to Theories of International Conflict.* San Francisco: Jossey-Bass.

Civic, Melanne Andromecca. 1998. A New Conceptual Framework for Jordan River Basin Management: A Proposal for a Trusteeship Commission. *Colorado Journal of International Law and Policy* Vol. 9 (Summer): 285.

Davis, David Howard. 1999. Great Lakes Commentary: Water Diversion from the Great Lakes. *The Toledo Journal of Great Lakes' Law, Science & Policy* (fall 1999): 109.

Hardin, Garrett. 1968. The Tragedy of the Commons. *Science,* December, 1243–8.

Editorial. 2001. Maybe More River Peace is Possible. *Omaha World-Herald* (Omaha, Nebraska), 12 May.

Seabrook, Charles. 1998. The Chattahoochee: River in Peril; Midwestern Omen? *The Atlanta Constitution (Atlanta, Georgia),* 5 October.

Six
Denver Water Politics, Two Forks, and Its Implications for Development on the Great Plains

Brian A. Ellison

In *Nature's Metropolis*, William Cronon chronicled the relationship between Chicago and the West. Chicago's ascendancy, he asserted, was intricately tied to the economic linkages it created with the agriculture, ranching, and transportation economies of rural America (Cronon, 1991). Thus, like a great sponge, Chicago absorbed, moved, and stored the economic booty that flowed from the farm towns of Iowa and the Great Plains beyond. Of course the same economic relationships between urban and rural that Cronon explored mirror those of other cities that ring and draw their livelihoods from the Great Plains, such as Kansas City, Tulsa, Rapid City, and others.

Denver, Colorado, however, offers a bit of a different story. Although Denver lies on the western edge of the Great Plains, its economy has been built on a more diversified base than that area could provide. Denver started as a mining camp at the foothills of the Rocky Mountains, and its most prolific source of wealth was based on the extraction of natural resources to the west. For the most part, Denver was built to supply the raw materials needed for development —expertise in finance, banking, law, transportation, and real estate— and to move west beyond the Great Plains. This is not to say that Denver's growth and development have not been tied to the Great Plains economy—indeed, many still consider it a "cow town"—but the city has looked beyond the ranching and agriculture economies that dominate its eastern flank for its vibrancy (Dorsett, 1977; Ellison, 1993; Leonard and Noel, 1990).

Why is Denver different? Unlike other cities on the Great Plains, Denver is not a natural transportation hub. Thus, while Chicago's access to the Great Lakes and Kansas City's access to the Missouri River provide a transportation rationale for the construction of those cities, it was extremely difficult to travel over the Rocky Mountains from Denver in the late 1800s. Denver was not in the path of the Transcontinental Railroad in 1869 and instead had to make do with a line that connected it to the main route to the north. Almost ninety years later Denver's leaders faced the same problem with the construction of the interstate highway system—a system that would have avoided the costly construction of I-70 between Grand Junction, Colorado, and Denver without the intervention and support of President Eisenhower (Dorsett, 1977; Ellison, 1993; Leonard and Noel, 1990).

Another variable that helps explain the difference between Denver and other cities of the Great Plains is its lack of water resources. Denver was not built in an ideal location for a large city. The region's groundwater is nonrenewable; the Rocky Mountains block most of the moisture from the west; and its natural sources of water, Cherry Creek and the South Platte River, were inadequate by the 1870s (Ellison, 1995, 1998). Once again, this is not to say that other Plains cities do not have water problems, but that the scope of Denver's water challenge was extraordinary.

Thus, faced with poor location and a lack of water, Denver's leaders have been forced to scramble for survival. More than any other variable, Denver's ascendancy has been based on politics: the flat-out ability of the city's leaders to make hard choices, avoid entanglement with the federal government, and make the most of Colorado's legal institutions to achieve their objectives. And nowhere, even with the construction of Denver International Airport, the crown jewel of Denver as transportation crossroads, are these political proclivities more evident than in the area of water resources development. Water is the key to understanding Denver's past and its future.

PLAN OF THE CHAPTER

Denver's economy is the glue that binds a vast area of the Great Plains to the Rocky Mountain west. To that end, Denver's struggle with water resources development can serve as a model for the region. This chapter examines the politics of water resources development in

Denver, focusing especially on the Two Forks water controversy of the late 1980s, with an eye toward drawing lessons for the region from that experience. A brief introduction to the politics of water resources development will be provided in the following section. Next, the chapter will turn to the Denver case study, focusing specifically on Two Forks. Finally, lessons from the case study will be assessed and generally applied to the Great Plains.

THE POLITICS OF WATER RESOURCES DEVELOPMENT

Perhaps the fundamental goal of policy scholars is to explain how values are distributed in American politics, recognizing that the most important parts of the policy process are informal. In other words, it would be extraordinarily naïve to think that the American policy process could be understood by examining its formal components, by reading the Constitution and knowing the mechanics of how a bill becomes a law. The truth is that the policy process in America has as much to do with informality—the committee system, alliances, log-rolling, and so on—than anything else. Indeed, scholars such as Deborah Stone (1997), contend that political decisionmaking is an art built on loyalty, persuasion, and passion—variables difficult to measure, quantify, or understand in a formal context. The broad question policy scholars ask is how we should organize and combine the formal and informal processes of American politics in order to understand how policies are made.

Most policy scholarship has been built on the typology first introduced by Theodore Lowi in 1964, and subsequently modified by many scholars (Greenberg et al., 1977). Lowi contended that American public policymaking could be organized into three primary categories: regulatory, redistributive, and distributive. The politics of regulatory policy, such as environmental legislation, is highly pluralistic and conflict-ridden. Interest groups from across the spectrum participate in regulatory politics, and issues are often debated in public. Redistributive policy, involving issues such as welfare and taxation, also creates highly conflict-ridden politics, but elites such as members of Congress and agency officials tend to dominate debate about these issues.

The politics surrounding distributive policy also tend to be dominated by elites, but conflict is low. The goal of distributive policymaking is to get taxpayers from across the country to pay for particularized benefits for special constituencies. Logrolling keeps system going—"I'll scratch your back if you scratch mine"—as members of Congress exchange votes on water projects, agricultural subsidies, weapons spending, and so on. Distributive policies produce what some deride as "pork-barrel spending," and others champion as infrastructure (Lowi, 1979). Distributive policies are controlled by tripartite alliances between congresspersons and their committees, executive agencies, and local beneficiaries of government spending (deHaven-Smith and Van Horn, 1984; Freeman, 1965; Hamm, 1983; Ripley and Franklin, 1984). These alliances are called "subgovernments," but over the years have also been called "whirlpools" (Griffith, 1939), "iron triangles" (Cater, 1964), and "cozy triangles" (Davidson, 1977).

Some scholars contend that policy alliances are larger and more diverse than described by the subgovernment concept (Baumgartner and Jones, 1991; Jenkins-Smith and Sabatier, 1994). Heclo (1978) argued that journalists, academics, environmentalists, and other interested citizens and groups have a role to play in broader "issue networks." Sabatier called these alliances "policy subsystems" and contended that they are composed of "journalists, analysts, researchers, and others who play important roles in the generation, dissemination, and evaluation of policy ideas" (Sabatier, 1987, 652). Rather than forming subgovernments, "advocacy coalitions" emerge among groups within subsystems to compete for control of policy outputs (Jenkins-Smith and Sabatier, 1994).

The politics of western water resources development has long been a favorite arena of study for scholars interested in distributive policymaking, subgovernments, and subsystems. The success of the water resources development subgovernment composed of congresspersons from western states, the Bureau of Reclamation, and local project beneficiaries has been the primary reason for this interest (McCool, 1994; Miller, 1985). By the mid-1980s, the Bureau of Reclamation had constructed hundreds of storage reservoirs and diversion dams, thousands of miles of canals, tunnels, pipelines, and laterals, while generating billions of dollars in agricultural, municipal, industrial, and energy benefits (Hughes, 1986; Reisner, 1993; Wahl, 1989; Wiley and Gottlieb, 1982). Although the hegemony of the water development subgovernment went mostly unchallenged between the passage of the Newlands Act in 1902 and the early 1970s, evidence that regulatory interests were affecting the preferred choices of distributive policy makers began to emerge in the 1950s (Ingram, 1972; 1990). In 1955, environmentalists won a hard-fought victory when they saved Dinosaur National Monument by forcing the Bureau of Reclamation and its congressional allies to remove Echo Park Dam from the Colorado River Storage Project Act. And in the 1960s, environmentalists won a similar battle when Bridge Canyon Dam was removed from the Colorado River Basin Project Act (Mann, 1975). Similarly, although Jimmy Carter tried and failed to eliminate several pork-barrel water projects in 1977, his efforts did introduce reforms that improved the "process by which new starts were initiated . . . in terms of cost sharing, repayment, interest rates, and other issues of project merit" (Miller, 1985, 405). Moreover, changes brought to the policy subsystem by the environmental movement and its legislative successes, such as the National Environmental Policy Act, Clean Water Act, and the Endangered Species Act, have given groups with regulatory goals a firm role to play in water resource development decisions (Miller, 1985).

McCool (1994) has questioned the political significance of these changes, suggesting that the distributive water subgovernment has simply learned to hand out "green pork" (such as money for biological and habitat studies or legal fees for environmental activist groups) to regulatory interests. However, there have been real changes in substantive policy outputs. Since the early 1980s, Congress has had a difficult time appropriating funds for project construction (Beard, 1993; U.S. General Accounting Office, 1983). Moreover, the agencies that once spent their time constructing water projects for the constituents of well-placed members of Congress now perform the myriad of duties needed to keep authorized projects alive, such as updating environmental impact statements, performing due diligence on water rights, recomputing cost-benefit analyses, updating compliance with the Endangered Species Act, and so on. Today, most substantive gains in water resource development are measured by the successful compliance with administrative procedure rather than the actual construction of water projects (Ellison, 1999).

To determine the extent to which these gains in water resource development can be credited to Denver and its water politics is one purpose of this chapter. To what extent have the water developers in Denver been able to achieve their goals? Has the policy environment changed over the years, and if so, how and why did it change? Have regulatory interests, such as environmentalists, been able to make inroads in the distributive decisionmaking system?

DENVER WATER POLITICS

In order to understand Denver water politics, it is critical to examine two fundamental issues. First, because Denver's development and growth have always been dependent on the availability of water, the city's founders built a water development institution with extraordinary power and ability. The Denver Board of Water Commissioners controls most of the formal water development mechanisms in the region and lies at the heart of Denver's economic development success. Second, Denver's phenomenal growth has been accompanied by regional growth and development. Thus, as Denver has grown, other county and local governments have also grown, in number and in power. Because, as in Denver, water is the key to their development, these governments compete with the Denver Board of Water Commissioners for limited water resource development opportunities. It is these two tensions that explain Denver water politics.

Tension and Competition in Denver Water Politics

The city and county of Denver were created in 1902 when state voters amended the Colorado Constitution to consolidate parts of three counties and the city into a single government. At the time, Denver was a multicounty city with urban problems that state and county officials felt should be managed and funded by the citizens of Denver. Thus, state voters also gave Denver strong annexation powers in order to avoid future intergovernmental conflicts. Areas of surrounding unincorporated counties could be added to Denver with the approval of the city council and the registered voters within the boundaries of the annexation (Dorsett, 1977; Ellison, 1993).

Because water was the key to the city's growth and development, Denver's citizens purchased the Denver Union Water Company and created the Denver Board of Water Commissioners in 1918. Because Denver's progressive leaders wanted the water department to be "managed by a nonpolitical board of businessmen," the new water department was organized along the commission model (Pickering, 1978). Members of the board were appointed by the mayor, served six-year overlapping terms, and could not be removed from office. Additionally, the new water commissioners were given complete control over the water department's revenues, which were placed in a special water fund separate from the city's general fund. By separating the board from the political influences of elections and the annual budget process, the progressives hoped that the commissioners would be free to apply business principles to management of the water utility. Although the water commissioners need voter approval before projects can be financed with municipal bonds, their one source of accountability, public appeals for funds have rarely been denied. In the face of board reports predicting higher costs, more conservation, and limited growth, most citizens eagerly approved new projects (Johnson, 1969; Kahrl, 1982; Miller, 1971).

The Denver Water Board competes with two sets of functional rivals in pursuit of its water development mission. As a water resources development agency—building large diversion, storage, and treatment facilities—the board competes with water developers on both sides of the Rocky Mountains for limited project construction opportunities. The Bureau of Reclamation, for example, has sponsored several massive irrigation projects on both sides of the mountains. One Bureau client in the area, the Northern Colorado Water Conservancy District, currently operates the largest transbasin water diversion project in the world (Tyler, 1992). Additionally, several cities, including Englewood, Thornton, and Colorado Springs, have developed their own water projects. Although this activity does not directly affect the board's ability to develop its water rights, which are protected by the Colorado Constitution, it does exacerbate statewide tensions that make development more difficult.

Next, as a water utility—delivering clean water to the citizens of Denver and leasing water to area inhabitants—the Denver Water

Board competes with other municipal water departments in the metropolitan area. During its early years, the board dominated competition for new service areas because it controlled most of the water. In the 1950s, the board imposed several service restrictions on the suburbs that forced many of them to develop independent water supplies. By the 1970s, these independent water departments were positioned to offer service to new areas. Suddenly, suburban politicians saw revenue opportunities in the unincorporated developments that were springing up around Denver. Recognizing Denver's past annexation successes, these politicians promoted plans that would allow their general-purpose governments to extend services to new areas, annex them, and collect their sales and property taxes (Ellison, 1993).

Standing between regional and suburban plans for water resources development are several institutional arrangements that make the Denver Water Board a formidable obstacle. The doctrine of prior appropriation, the fundamental tenet of Colorado water law, gives the Denver Water Board an advantage in the competition because its first members—the board itself along with the banking, construction, and business communities that had the most to gain from Denver's growth—had the foresight to secure the western slope water rights needed to serve what some hoped would become a 600-square mile supercity (Miller, 1971; Milliken, 1988).

Additionally, because Denver's progressive leaders feared the chicanery of state politicians, Denver's home rule amendment included provisions that protected its right to operate a municipal utility (Colorado Constitution, art. 20, sec. 6). Article 5, section 35 of the Colorado Constitution states that the "General Assembly shall not delegate to any special commission . . . any power to make, supervise, or interfere with any municipal improvement, money, property, or effects . . . or perform a municipal function whatever." Article 25 established the Colorado Public Utilities Commission, whose members are appointed by the General Assembly, and empowered it to regulate utilities "provided, however, nothing herein shall affect the power of municipalities to exercise reasonable police and licensing powers, nor their power to grant franchises; and provided, further, that nothing herein shall be construed to apply to any municipally owned utilities."

What makes these institutional arrangements troublesome for the Denver Water Board's competitors—other water developers and municipal utilities—is that these provisions in the Colorado Constitution also protect them. Elected officials and government managers in other suburban cities also want to develop their water rights and water utilities but do not want the state's Public Utilities Commission to regulate their activities. Thus, any changes in these arrangements that would hinder the Denver Water Board would also affect them.

The Board's Allies

To fully appreciate the challenge of Two Forks, it is important to review the Denver Water Board's relationship with the governments that lease its water. These entities—mostly special district governments and the region's independent water developers and municipal utilities—play a special role in the intergovernmental mix that dominates water politics in Denver.

Many of the people moving into the booming Denver metropolitan area during the 1950s and 1960s decided to live outside the city. In 1940, 28% of the metropolitan population lived outside Denver. By 1960, that number climbed to 49% of the total metro population; by 1986 it was 75%. Meanwhile, Denver's population had stabilized at approximately 500,000 people by 1960 (Ellison, 1993).

Because these people were moving into the unincorporated developments outside Denver, they were without public services and understandably turned to the water commissioners for water service. In response to this demand, the board implemented several policies designed to protect water quality and the structural integrity of their water system. First, the board signed distribution contracts with special district or municipal governments only, reasoning that these governments would have the taxing authority to maintain their water systems. Second, the board enforced operational compliance through its *Engineering Standards* manual, and its rules and regulations for water distribution. Taken together, these actions helped assure the board's engineers that these systems would be in compliance with their rigorous technical standards when the systems were acquired through annexation sometime in the future (Ellison, 1993).

Two things happened, however, that created a series of unusual institutional arrangements in Denver water politics. First, the Denver Water Board's operational policies created a proliferation of special district governments in the Denver metropolitan area, specifically created to distribute its water. Second, by the early 1970s, suburban residents decided that they no longer wanted to live with the threat of Denver's aggressive annexation posture and successfully ended that threat by amending the state constitution in 1974. The Poundstone Amendment required that Denver city council seek the approval of all county voters before unincorporated areas in their territory could be annexed. Nevertheless, just as regional governments felt freed from Denver's annexation threat, in their midst were dozens of special district governments with institutional ties to the Denver Water Board. Instead of city officials from Littleton, Colorado, governing the municipality's most precious resource, water service was still controlled by the Platte Canyon Water and Sanitation District and the Denver Water Board. Thus, even though the relationship between the city of Denver and the suburbs had been radically altered, the fundamental institutional relationships that controlled growth and land-use planning were still in place (Ellison, 1993; 1995).

Foothills to Two Forks

Because it was clear that areas outside the city would need water service, the Denver Water Board was given an intergovernmental mission: to ensure an adequate supply of high-quality water to Denver and its inhabitants and to lease water not needed by Denver to the surrounding metropolitan area (Miller, 1971). In practice, the water commissioners have made a firm distinction between Denver and suburban customers, for example, by supporting almost unlimited development inside Denver, charging Denver customers lower rates, and imposing service restrictions on suburban customers first during water shortages (Ellison, 1993; 1995; 1998).

Although Denver's population had basically stabilized by the 1960s, especially in proportion to the water department's capacity to serve Denver residents, the water commissioners continued to expand the department's capacity to provide water service to the suburbs. During the 1960s, for example, with completion of their Blue River Diversion Project, the water department doubled its inflow and storage capacity. By 1973, facing extreme shortages in treated water supply, the board proposed the construction of a new, massive water-treatment plant on its property in the foothills southwest of Denver (Ellison, 1993).

The water commissioners used suburban demand for treated water to justify construction of the Foothills Water Treatment Plant. This justification, however, proved costly for the board and ultimately led to major battle. Denver residents did not understand why they should finance a suburban water project and refused to pass the board's first bond issue. Although the board received support for the project from its special district water suppliers, the suburban general-purpose governments did not want to be further tied to Denver's apron strings.

In response, the water commissioners resorted to tested tactics to create support for their project. First, the board imposed a moratorium on new suburban service, which moved Denver's voters to approve the agency's funding referendum but further antagonized the suburban general-purpose governments. Second, after years of delay caused by the annexation disputes and effective court challenges brought by environmentalists, the board restricted lawn and garden watering, and implemented a tap-allocation program. Finally, the pressure to resolve the Foothills crisis led to the development of an arbitrated settlement. In exchange for the permits needed to build the project, the board agreed to implement a water conservation program, improve fish habitat, create a forum for public participation, and pay the legal expenses of several environmental interest groups (Ellison, 1993).

Although the metropolitan area was reeling from the Poundstone Amendment and the Foothills compromise, Denver and suburban politicians promoted metropolitan cooperation as the solution to the region's problems. The metropolitan cooperation concept was first introduced as a solution to the annexation crisis of the 1970s. At the heart of the metropolitan cooperation discussion was the argument that problems between governments could be overcome by sharing resources. The suburban general-purpose governments, for example, wanted area governments to combine their water resources to form a metropolitan water authority. Denver wanted assistance with services and amenities that the city's residents paid for but suburban residents enjoyed, such as hospitals, the airport, parks, and cultural facilities (Davis, 1989; Milliken, Lohman, and Gougeon, 1985).

The water commissioners also came to realize in the early 1980s that they would no longer be able to single-handedly achieve their water development goals in the face of extreme opposition from environmental groups, federal agencies, elected officials, and the suburban general-purpose governments. In order to continue expanding its system, the board would need new allies in the spirit of metropolitan cooperation to help balance the overwhelming pressures against development. In 1982, the board turned to the one group that had supported the Foothills project, but that had been swamped in the ruckus of the conflict—their special district government water distributors. Thus, in July 1982, the board signed an agreement with forty-two water distributors that would allow them to share the costs and benefits of new water developments. Specifically, the Metropolitan Water Development Agreement (MWDA) split the costs and benefits of all future projects at 20% for the Denver Water Board and 80% for the distributors. In 1984, concerned about rapid population increases in the Denver area, the water commissioners signed yet another contract that designated the objective of their partnership with the special district governments: the Two Forks Dam and Reservoir and the Colorado River Joint Use Reservoir (Denver Board of Water Commissioners, 1984; Ellison, 1993).

Two Forks

Two Forks Reservoir had a long history at the water department. The water rights that would be used to fill the reservoir were first claimed in 1931 on the Blue River. During the late 1940s and early 1950s, the Bureau of Reclamation hoped to build a structure on the confluence of the North Fork and South Platte Rivers that would hold water for farmers and municipalities. Next, in 1972, the Board announced plans to build Two Forks Reservoir in conjunction with the Foothills Treatment Plant and the Strontia Springs Reservoir, but the plan was rejected when Denver's residents failed to pass the bond issue.

After it became apparent that the citizens of Denver did not want to support a massive water project for the suburbs, the Board settled for the Foothills complex alone. But the environmental community was not fooled by the board's change of heart. Foothills was built to support massive inflows—greater than the South Platte's current yield—from a large reservoir. The Foothills Water Treatment Plant, for example, was designed to accommodate 125-million-gallon-per-day increases in its treatment capacity—up to 500 million gallons per day. This meant that the Board anticipated more water development projects on either the South Platte or the western slope. Additionally, Strontia Springs Reservoir provided a more obvious clue because it would make a perfect afterbay for a larger power-generating reservoir.

Environmentalists criticized Two Forks because it would inundate 13.5 miles of free-flowing river and destroy habitat with its 1.1-million-acre-foot capacity. Moreover, environmentalists claimed that the water department had not fully implemented the conservation and metering provisions stipulated in the Foothills compromise. Federal officials were concerned with the project because the board seemed to be violating the spirit of the National Environmental Policy Act—especially the search for alternatives to development.

In 1984, the Board agreed not to apply for the dredge-and-fill permits it needed to build Two Forks until the systemwide environmental-impact statement (SEIS) it had agreed to write as part of the Foothills compromise was completed. In exchange, it wanted the Army Corps of Engineers to include a site-specific environmental-impact statement for Two Forks. Although the Corps complied with the board's request, costs for the environmental-impact statement were quickly approaching $30 million, and both the commissioners and their water distributors had run out of patience. In 1986, the board filed for the permits needed to build Two Forks.

Finally, in 1987, the Army Corps of Engineers released the draft SEIS. The U.S. Environmental Protection Agency (EPA), charged with reviewing the draft and final environmental impact statements, reacted to the Corps' document with dismay. EPA charged that the Corps had selectively withheld information from the environmental-impact statement, and that its discussion of the alternative projects, conservation measures, and mitigation plans were inadequate. Moreover, the EPA accused the Corps of shoddy work, and of neglecting important issues like air and water quality (Ellison, 1993). The EPA also wanted the Corps to seriously consider several alternatives, such as enlargement of Cheesman Reservoir, a large and small Eastabrook Dam on the North Fork, and a smaller, around 400,000 acre-feet, Two Forks reservoir. Other groups also criticized the findings of the SEIS. Environmental interest groups, such as the Environmental Defense Fund and the Colorado Environmental Caucus, charged that the estimated safe annual yield from Two Forks, 98,000 acre-feet per year, was hydrologically inefficient (Ellison, 1993; Leucke, 1990).

Meanwhile, as the Corps began work on the final SEIS, the Board worked fast to develop an extensive mitigation program for Two Forks. The Board estimated that mitigation would cost $45 million when they made the plan public in 1987, and it encouraged the U.S. Fish and Wildlife Service to endorse Two Forks that same year. By mid-1988, however, many environmentalists charged that the Board had used undue influence on the Fish and Wildlife Service to get the recommendation it wanted.

In March 1988, the Corps released the final SEIS, and extended the public comment period on the project to June 10. Additionally, the Corps announced that a final permit for the project would be granted pending the outcome of the public hearings. On April 12, in reaction to threats against their project, the Board released a position paper that predicted what would happen if the permit was not granted: the imposition of more stringent conservation measures, including mandatory restrictions on the size of lawns for new homes; increased use of nonrenewable groundwater; a renewed pursuit of agricultural water rights by metropolitan governments in northern Colorado; and a reduction in the incentives for metropolitan cooperation (Ellison, 1993).

As the Board worked to increase public support for Two Forks, the environmental community was launching its own attack on the project. This time, however, the Board's opponents not only focused on the project's devastating environmental effects, but also worked to undermine the economic and technical justifications for the project. Dan Luecke of the Environmental Defense Fund, for example, called the project "financial suicide," noting that the project's purported $440 million price would probably be closer to $4.1 billion. Moreover, the environmental coalition's principal players—the Environmental Defense Fund, Trout Unlimited, and the National Audubon Society—were able to evaluate and challenge the project's technical merits. Their evaluation, combined with opposition from lower-level government officials, allowed environmentalists to undermine public support for the project (Ellison, 1993).

In May of 1988, despite recommendations by the Corps, the EPA announced that it was opposed to the immediate construction of the reservoir. This announcement gave Colorado Governor Roy Romer the room he needed to introduce a compromise solution to the project. In June, Romer recommended that the Corps grant the permits needed to build the project with a twenty-five-year shelf life. This way, Romer argued, the Board could wait and see if its demand projections were accurate before beginning construction. But Romer's recommendation, although appealing to many government officials and citizens that wanted a cautious approach to development, proved to be a disaster for the Board. The water commissioners had been selling the project based on statistical projections that predicted immediate threats to the city's water supply; thus Romer's proposal meant that he did not accept the Board's arguments.

By mid-1988, public support for the project was decreasing as the pressure on the Corps to issue a final decision was increasing. Many Two Forks proponents, such as the Denver Regional Council of Governments, the Colorado Homebuilders Association, and labor groups, heightened their call for a permit. And agreeing with Governor Romer, Denver Mayor Federico Peña gave the project his approval, claiming that a permit with a twenty-five-year shelf life would give the city flexibility in meeting water demands.

Finally, in January 1989, the Corps announced that it would grant the permits needed to build Two Forks if the board agreed to comply with an impressive list of conditions. But although the water commissioners were relieved, their partners—the special district water distributors—accused the Corps of going beyond its legal boundaries in its permit conditions. Ultimately, however, the distributors' efforts to reduce the project's cost were futile, as the EPA announced in March 1989 that it would use its authority under the Clean Water Act to veto the Corps's decision. Despite their cautious support for the project, the EPA's actions infuriated Governor Romer, Mayor Peña, and Colorado's congressional delegation.

By early 1990, the EPA had taken the necessary steps to officially veto the Two Forks project. The Board had expected a fight over Two Forks, but they also expected to win if they played according to the rules. The Board followed the procedures stipulated by the National Environmental Policy Act, the Clean Water Act, and a variety of other federal statutes, and after agreeing to pay for the costs of mitigation, the Corps agreed to issues the necessary permits. Although the process had cost over $40 million, producing the largest and most costly environmental-impact statement ever, the Board felt that the system had worked. The environmental-impact statement process, they argued, was not supposed to stop projects that legitimate local agencies needed. Instead, it was intended to be a planning mechanism, designed to bring environmental considerations into the process of developing resources (Ellison, 1993).

For many local and state officials, however, Two Forks represented more than just another water project. Two Forks was an attempt to reform local government, to create new institutional structures that would instigate metropolitan cooperation on a variety of fronts. By 1989, for example, fourteen general-purpose governments had joined the special districts in their agreement to build Two Forks with the Board, and it seemed that the formation of a metropolitan water agency was inevitable. But as predicted, by the time EPA issued its formal veto of the project in 1990, the uneasy coalition of water developers had fallen apart.

After Two Forks: 1990–1992

Shortly after the Two Forks veto, the Board announced a variety of new programs and policies that the suburbs claimed would shift the burden of impending water shortages to them. These new policies included a continuation of the out-city tap allocation program, restrictions on new service areas, and an increase in out-city rates (Denver Board of Water Commissioners, 1989). Additionally, the Board reversed several of the policies it held with regard to conservation. Among the most important of these was the imposition of a progressive rate structure, which increased the price of water with consumption, rather than decreasing it. The Board also stepped up its efforts to implement universal metering, and it introduced a cash rebate program for customers that installed ultralow-volume toilets.

These actions infuriated the Board's water distributors, who claimed that the agency demonstrated a "Jekyll and Hyde" attitude, "expecting and getting cooperation when it needs help on issues . . . but then refusing to extend the same spirit of cooperation to water policy" (Massaro, 1989, 8). The Board distributors hoped their partnership was perpetual—that despite the Two Forks veto they had become partners with the water commissioners. But the Board was constrained by the Denver Charter, especially the provisions that governed water sales to entities outside the city. By May of 1990, metropolitan cooperation was in serious trouble because several distributors, under their new organization, the Metropolitan Denver

Water Authority, launched a lawsuit against the EPA (*Alameda Water & Sanitation District et al. v. U.S. Environmental Protection Agency.* Civil Action No. 91-M-2047. U.S. District Court, District of Colorado, 1992a).

After the EPA announced its intention to veto Two Forks, the Board reaffirmed its commitment to the project, stating that it was "determined to continue to protect the water rights and prior approvals for the Two Forks project because the Board was convinced that... Two Forks is ... far and away the best water program for the Denver metropolitan area" (Denver Board of Water Commissioners, 1989). And for a brief period it seemed that Two Forks was far from dead as EPA officials hinted that a smaller dam and reservoir might still be possible. But while the proponents and opponents of Two Forks geared up for a second battle, Wellington Webb was running for mayor of Denver with an anti-Two Forks plank in his platform.

After Webb was elected, the Board shifted its position on Two Forks. Although the Board convinced the mayor that abandoning the project was a mistake—the water rights, for example, were extremely valuable, not to mention the dollars spent on the SEIS—the agency decided not to join the distributors in the lawsuit against the EPA (Ellison, 1993). The Board's refusal to join the lawsuit ended the metropolitan cooperation initiative. After all, it was the Board's permit that was denied, not the water distributors'. Thus, the EPA quickly filed a motion for dismissal claiming that the distributors lacked standing to sue (*Alameda Water & Sanitation District et al. v. U.S. Environmental Protection Agency.* Civil Action No. 91-M-2047. Reply Memorandum in Support of Defendants' Motion to Dismiss for Lack of Standing, 1992).

In June 1992, the Board announced that it was suspending work on Two Forks and pursuing a policy of retrenchment regarding water service to suburban governments. The Board is still pursuing new sources of raw water, but is focusing on water exchanges and expansions of current facilities. Perhaps most interestingly, the Board is interested, like other metropolitan cities, in water exchanges and purchases from agricultural suppliers. Although there are some reservations about such exchanges among agency employees, these projects have the capacity to provide Denver with water well into the twenty-first century.

DISCUSSION

As with the federal water resources development model, water politics in Denver was dominated by a distributive subgovernment. As documented in other cities (Domhoff, 1986; Stone, 1989), bankers, realtors, developers, and utilities worked together to ensure that growth would continue along with their financial fortunes. The Denver Board of Water Commissioners is the centerpiece of this growth subgovernment. The relationship between the Board and other progrowth components of the subsystem is revealed through an examination of the water commissioners and their backgrounds. Between 1918 and 1992, there were fifty-three water commissioners. Thirty-seven of them were in business, either banking, real estate, or commerce, and twenty-two had some experience in Denver politics. Seven of the commissioners were engineers, and thirteen of them were attorneys. Only three commissioners were women, and only one of the commissioners could be labeled an environmentalist (Ellison, 1993; 1998).

Both the state of Colorado and the federal government were original members of the water development subgovernment. Colorado, for example, has participated and struggled to enhance expansion of the Denver metropolitan area through the Colorado Water Conservation Board, which was created in 1937 "for the purpose of aiding in the protection and development of the waters of the state" (Colorado Revised Statutes, 37-06-102). A host of federal agencies has worked to support water resources development in Colorado and for the Denver metropolitan area. The Bureau of Reclamation, for example, conducted extensive water-use studies on the Blue River during the 1930s and 1940s and actively encouraged the construction of projects to move water over the Rocky Mountains to Denver (U.S. Department of the Interior, 1948).

Perhaps the first challenge to Denver's dominant position in water resource development came from the suburban and county governments outside the city. Although the counties originally supported Denver, they now work against Denver to support their own growth. Adams, Jefferson, and Arapahoe Counties have active economic development programs and work ardently to secure new sources of revenue from expanding tax bases. Suburban municipalities also support metropolitan growth and are now working to develop their own water resources that can be used to encourage development. Cities such as Aurora and Thornton have spent millions of dollars on legal expenses associated with the transfer and purchase of agricultural water rights (Ellison, 1993).

Moreover, by the late 1960s and 1970s, transportation problems, chronic air pollution, intergovernmental rivalry for new sources of tax revenues, environmental degradation, sprawl, and a deteriorating quality of life triggered a debate about the costs, direction, and even the philosophy of growth. Federal environmental protection legislation also gave environmental and antigrowth advocates a forum for participation in regional planning processes. These regulatory changes have had a significant impact on the subgovernment, as groups with fundamentally different ideas about the nature of growth have entered the debate and now compete with traditional progrowth coalitions. New members of what can now be called Denver's growth subsystem, including an advocacy coalition composed of the Environmental Protection Agency, the Colorado Division of Wildlife, the Environmental Defense Fund, and others, actively question the idea that growth should continue as long as new sources of water can be provided.

Federal changes in water-resources management policy also forced the water commissioners to pursue goals contrary to their agency's fundamental mission, which can be documented by examining agency participation in the Two Forks permitting process. In order to acquire the Clean Water Act dredge-and-fill permits required to build the project, the Denver Water Board and its allies had to examine a variety of alternatives and pursue unique goals. Many environmental organizations, for example, argued that the water department had not considered conservation as a source of supply in its hydrological models. But the commissioners and the water department's engineers were hesitant to pursue regulatory programs that might fail during droughts (Ellison, 1993). Moreover, the water commissioners had to work with a large number of groups with access to the permitting process, including ten federal, thirteen state, and four local agencies; the Denver Regional Council of Governments and the Northwest Colorado Council of Governments; Grand, Douglas,

Summit, and Jefferson Counties; the State of Nebraska; and a host of environmental interest groups (Ellison, 1993).

LESSONS FOR THE GREAT PLAINS

What does the Two Forks experience mean for the Great Plains? First, traditional developmental subgovernments—whether they are built on the need to develop water resources, graze cattle on public land, or secure protracted agricultural subsidies from the federal government—probably no longer exist. If they do exist, they now share policy space with a variety of other interests and groups with substantive political prowess and expertise. Just as the water commissioners and their allies can no longer dictate the future of water resources development in the Denver metropolitan area, it is unlikely that agricultural and ranching interests can dictate the future of the Great Plains.

Second, and perhaps more importantly, the development of natural resources is now an intergovernmental task. Any development decision will involve federal, state, and local agencies working together to implement policies created by all levels of government. Moreover, because implementation of diverse policies for developing natural resources, protecting endangered species, creating alternatives to development, protecting water resources, and so on, requires input from the public, concerned interest groups, and various governments, cooperation must be at the center of any activity. Development will no longer occur unless agencies work to address diverse interests and the goals of many groups.

REFERENCES

Baumgartner, F., and B. D. Jones. 1991. Agenda Dynamics and Policy Subsystems. *Journal of Politics*. 53: 1044–74.

Beard, D. P. 1993. Blueprint for Reform: The Commissioner's Plan for Reinventing Reclamation. Washington, DC: U.S. Bureau of Reclamation.

Cater, D. 1964. *Power in Washington.* New York: Random House.

Cronon, W. 1991. *Nature's Metropolis: Chicago and the Great West.* New York: W. W. Norton.

Davidson, R. H. 1977. Breaking Up Those 'Cozy Triangles': An Impossible Dream? In *Legislative Reform and Public Policy.* S. Welch and J. G. Peters (eds.). New York: Praeger.

Davis, S. K. 1989. Preliminary Reconnaissance: Metropolitan Denver Cooperation on Water. Fort Collins, CO: Colorado Water Resources Research Institute.

deHaven-Smith, L., and C. Van Horn. 1984. Subgovernment Conflict in Public Policy. *Policy Studies Journal.* 12: 627–42.

Denver Board of Water Commissioners. 1984. Platte and Colorado River Storage Projects Participation Agreement. Denver, CO: Central Records..

Denver Board of Water Commissioners. 1989. Water Conservation Plan of the Denver Water Department. Denver, CO: Office of Water Conservation..

Domhoff, W. G. 1986. The Growth Machine and the Power Elite: A Challenge to Pluralists and Marxists Alike. In *Community Power: Directions for Future Research.* R. J. Waste (ed.). Newbury Park, CA: Sage Publications.

Dorsett, L. W. 1977. *The Queen City: A History of Denver.* Boulder, CO: Pruett Publishing.

Ellison, B. A. 1993. The Denver Water Board: Bureaucratic Power and Autonomy in Local Natural Resource Agencies. Ph.D. dissertation, Colorado State University, Fort Collins, Colorado.

Ellison, B. A. 1995. Autonomy in Action: Bureaucratic Competition Among Functional Rivals in Denver Water Politics. *Policy Studies Review* 14, no. 1-2: 25–48.

Ellison, B. A. 1998. Intergovernmental Relations and the Advocacy Coalition Framework: The Operation of Federalism in Denver Water Politics. *Publius: The Journal of Federalism* 28, no. 4: 35–54.

Ellison, B. A. 1999. Environmental Management and the New Politics of Western Water: The Animas–La Plata Project and Implementation of the Endangered Species Act. *Environmental Management* 23, no. 4: 429–39.

Freeman, L. J. 1965. *The Political Process.* Rev. ed. New York: Random House.

Greenberg, G., J. Miller, L. Mohr, and B. Vladeck. 1977. Developing Public Policy Theory: Perspectives from Empirical Research. *American Political Science Review* 71: 1532–43.

Griffith, E. S. 1939. *The Impasse of Democracy.* New York: Harrison-Hilton Books.

Hamm, K. 1983. Patterns of Influence Among Committees, Agencies and Interest Groups. *Legislative Studies Quarterly* 8: 379–426.

Heclo, H. 1978. Issue Networks and the Executive Establishment. In *The New American Political System.* A. King (ed.). Washington, DC: American Enterprise Institute for Public Policy Research.

Hughes, S. 1986. Major Federal Water Resource Agencies: History and Functions of the Army Corps of Engineers and the Bureau of Reclamation. Washington, D.C.: Congressional Research Service.

Ingram, H. 1972. The Changing Decision Rules in the Politics of Water Development. *Water Resources Bulletin* 8: 1177–88.

Ingram, H. 1990. *Water Politics: Continuity and Change.* Albuquerque, NM: University of New Mexico Press, Albuquerque.

Jenkins-Smith, H., and P. A. Sabatier. 1994. Evaluating the Advocacy Coalition Framework. *Journal of Public Policy* 14: 175–203.

Johnson, C. A. 1969. *Denver's Mayor Speer.* Denver, CO: Green Mountain Press.

Kahrl, W. 1982. *Water and Power.* Berkeley, CA: University of California Press.

Leonard, S. J., and L. Noel. 1990. *Denver: Mining Camp to Metropolis.* Boulder, CO: University Press of Colorado.

Leucke, D. F. 1990. Controversy over Two Forks Dam. *Environment* 32: 42–45.

Lowi, T. J. 1964. American Business, Public Policy, Case Studies, and Political Theory. *World Politics* 16: 687–713.

Lowi, T. J. 1979. *The End of Liberalism: The Second Republic of the United States.* 2nd ed. New York: W. W. Norton.

Mann, D. E. 1975. Political Incentives in U.S. Water Policy: Relationships Between Distributive and Regulatory Politics. In *What Government Does.* M. Holden Jr. and D. L. Dresang (eds.). Beverly Hills, CA: Sage Publications.

Massaro, G. 1989. Suburbs Snipe at City Water Board. *Rocky Mountain News,* October 11, p. 8.

McCool, D. 1994. *Command of the Waters: Iron Triangles, Federal Water Development, and Indian Water.* Tucson, AZ: University of Arizona Press.

Miller, W. 1971. *The Denver Water Department. Genesis: from Privately-Owned Firms to Municipally-Owned Utility.* Denver, CO: Denver Board of Water Commissioners.

Miller, T. 1985. Recent Trends in Federal Resource Management: Are the 'Iron Triangles' in Retreat? *Policy Studies Review* 5: 395–412.

Milliken, J. G. 1988. Water Management Issues in the Denver, Colorado Urban Area. In *Water and Arid Lands of the Western United States.* M. T. El-Ashry and D. C. Gibbons (eds.). Cambridge, MA: Cambridge University Press.

Milliken, J. G., L. C. Lohman, and T. A. Gougeon. 1985. *Economic and Social Impacts on the City and County of Denver of Alternative Metropolitan Water Supply Policies, 1985-2010.* Denver, CO: Denver Research Institute.

Pickering, J. R. 1978. Blueprint of Power: The Public Career of Robert Speer in Denver, 1978-1918. Ph.D. dissertation, University of Denver, Denver, Colorado.

Reisner, M. 1993. *Cadillac Desert: The American West and Its Disappearing Water.* Rev. ed. New York: Penguin.

Ripley, R. B., and G. A. Franklin. 1984. *Congress, the Bureaucracy, and Public Policy.* 3rd ed. Homewood, IL: Dorsey Press.

Sabatier, P. 1987. Knowledge, Policy-Oriented Learning, and Policy Change. *Knowledge: Creation, Diffusion, Utilization.* 8: 649–92.

Stone, C. N. 1989. *Regime Policies: Governing Atlanta 1946-1999.* Lawrence: University Press of Kansas.

Stone, D. 1997. *Policy Paradox: The Art of Political Decision Making.* New York: W. W. Norton.

Tyler, D. 1992. *The Last Water Hole in the West: The Colorado-Big Thompson Project and the Northern Colorado Water Conservancy District.* Niwot, CO: University Press of Colorado.

Wahl, R. W. 1989. *Markets for Federal Water: Subsidies, Property Rights, and the Bureau of Reclamation.* Washington, DC: Resources for the Future.

Wiley, P., and R. Gottlieb. 1982. *Empires in the Sun: The Rise of the New American West.* New York: G. P. Putnam's Sons.

U.S. Department of the Interior, Bureau of Reclamation. 1948. Blue-South Platte Project Colorado: A Potential Transmountain Diversion Project. Denver, CO: Project Planning Report.

U.S. District Court, District of Colorado. 1992. *Alameda Water & Sanitation District et al. v. U.S. Environmental Protection Agency.* Civil Action No. 91-M-2047.

U.S. General Accounting Office. 1983. Water Project Construction Backlog—A Serious Problem with No Easy Solution. Washington, DC: GAO/RCED-83-49.

Seven

Federal Water Grants Participation: A Comparison of Arid States with Nonarid States

Charles R. Britton
Richard K. Ford

In *The Theme and Profile Investing: Special Report,* Merrill Lynch addressed the ever-present issue of clean water for the Millennium under the topic Environmental—Clean Water—Basic:

> Life depends upon clear, clean water. It is growing scarcer as the population grows, and water quality declines. The health-conscious population is demanding water purity and, increasingly there is industrial demand for clean water. Due to rising demand and increased cost of treating polluted water, the price paid for water is rising. The U.S. Environmental Protection Agency estimated that at least $138.4 billion must be invested to bring the nation's water systems up to Federal water quality standards. Privatization or outsourced operation of municipal water facilities is expected to grow substantially in coming years. (Merrill Lynch, 1998, 34)

ENVIRONMENTAL FEDERALISM

There seems little doubt concerning the veracity of the Merrill Lynch statement. The problem concerns the economic costs in the achievement of clean water. Here the problem centers on who sets the standards and who pays to meet them. The concept of environmental federalism is founded in the many different levels of government (federal, state, and local) in the United States. With respect to environmental objectives, however, the question must be posed concerning the rationale for assigning a superior value to federal standards over state and local standards. The charge is often made that, if states set environmental standards for clean air and water, there would be nothing but "bottom feeders" because all states attempt to set the lowest standards in an effort to attract industry and jobs. Many environmentalists and environmental economists contend that different locations have different economic costs and associated benefits and therefore should have different standards that would actually enhance economic efficiency.

> Curiously under the Clean Water Act . . . Congress has given the states the responsibility (but subject to EPA approval) for setting their own standards for water quality. (Oates, 1998, 15)

THE SAFE DRINKING WATER ACT

The Congressional Budget Office, in discussing the Safe Drinking Water Act (SDWA), passed by Congress in 1974, noted that it contained

> [t]he first set of federally enforceable standards for drinking water. . . . However, compliance with those standards was voluntary. The Congress recently amended the SDWA again. The 1996 amendments provide the EPA with more flexibility to consider costs and benefits in setting standards. (Congressional Budget Office, 1997, 17)

This chapter attempts to answer the question: What determines an individual state's desire to participate in SDWA and the EPA standards?

THE ECONOMIC MODEL

To analyze state participation in SDWA and voluntary compliance with Environmental Protection Agency (EPA) standards, we first collected data on individual states from the *1998 Budget of the United States Government*. We selected the "Budget Information for States" (BIS) that provided the actual obligations to states for fiscal year 1996 for different state and local government grant programs. We were able to glean the individual state data from the EPA's Office of Water programs: 1) Drinking Water State Revolving Fund (66.458), 2) Clean Water State Revolving Fund (66.458), and 3) Summary of Programs by State (six programs total including the first two listed).

We excluded Hawaii and Alaska to confine our analysis to the contiguous states. Next, we tried to determine the variables that would influence a state's propensity (as represented by its legislative body) to participate in the federal grants program. Being economists, we chose the typical variables: population, personal income, educational attainment, cost of living, and population density. We then added the following variables concerned with water availability and health: cancer rates, water area as a percent of total area, developed land area, per capita water use, total water consumption, and precipitation.

Preliminary Results

Each of the federal assisted water-related grant variables turned out to be statistically significantly related to the following independent variables:

1. total population
2. percent metro-population
3. total area
4. surface water area
5. federal lands
6. per capita water use
7. cancer death rates
8. poverty rates
9. population over 65 years

The total population variable was by far the most statistically important of all the independent variables. This of course was expected, and conformed to political considerations. One would expect that the distribution of federal monies somehow matched the population distribution because we live in a democracy where votes that represent political power are a function of the population distribution.

Arid/Semiarid-Humid/Semihumid Model

In an effort to account for the influence of the total population variable, each of the federal assisted water-related grant variables were adjusted (divided) by total state population. Then the grant variables were arranged from high participation to low participation and divided into three categories with an equal number of states (i.e., sixteen states per category). This yielded a grouping variable of the states for each of the three water-related grant variables. These forty-eight contiguous states were next categorized into two categories: 1) arid and semiarid (fourteen states) and 2) humid and semihumid (thirty-four states).

The criterion used for identification purposes was the generally accepted delineation mark of the 20-inch isohyet that runs through the Midwestern states. The 20-inch isohyet line extends north-south along the eastern portions of the states from North Dakota to Texas and roughly coincides with 100 degrees west longitude. The 20-inch isohyet is associated with evapotranspiration and represents the boundary between arid and humid regions. The areas to the west of the 20-inch isohyet line (less than 20 inches of annual precipitation) have potential evaporation and transpiration that exceed precipitation and are therefore classified as arid and semiarid. These western fourteen states include Arizona, Colorado, Idaho, Kansas, Montana, Nebraska, Nevada, New Mexico, North Dakota, Oklahoma, South Dakota, Texas, Utah, and Wyoming. Areas east of the 20-inch isohyet boundary (more than 20 inches of precipitation) receive more precipitation than is either evaporated or transpired and are generally viewed as having a surplus of water and are therefore classified as humid and semihumid. These thirty-four remaining states were classified as humid and semihumid (Britton, Ford, and Hehr, 1993).

Participation in Federal Water Grants

The participation of these states in the Federal Water Grants is indicated in Tables 7.1, 7.2, and 7.3. Table 7.1 presents the representation of the states in the Federally Administered Clean Water Revolving Fund (CWRF). Notice these states are ranked by their participation from low to high with one-third of the states falling in each category. Furthermore, note the dispersion of the number of states in the two columns: the Humid-Semihumid column has a relatively equal dispersion in the rows, while the Arid-Semiarid column has a distinctly bimodal distribution. Table 7.2 shows the results of the Drinking Water State Revolving Fund, and Table 7.3 displays the Summary of Water Related Programs, a total of six water-related programs including the two listed. If all the states participated evenly in the Federal Water Grants one would expect approximately eleven humid-semihumid states and five arid-semiarid states

119

in each category. However, it is obvious that this is not the case. This technique of data presentation is well suited to chi-square analysis.

Table 7.1.—Clean Water Revolving Fund

Participation Rate	Humid-Semihumid	Arid-Semiarid
Lowest 1/3	AR, VA, AL, CT, FL, OR, IL, PA (8)	CO, OK, TX, UT, NM, KS, NE, AZ (8)
Middle 1/3	IN, IA, MN, CA, MI, GA, NC, NJ, OH, LA, SC, WI, TN, RI, MS (15)	NV (1)
Highest 1/3	MA, KY, MO, ME, VT, WA, NH, MD, DE, NY, WV (11)	SD, ND, MT, ID, WY (5)
	Chi-square statistic 7.46 Probability of a greater chi-square statistic .024	

Table 7.2.—Drinking Water State Revolving Fund

Participation Rate	Humid-Semihumid	Arid-Semiarid
Lowest 1/3	MA, CA, TN, AL, FL, NY, IL, KY, MD, NJ, GA, OH, MO, SC (14)	TX, AZ (2)
Middle 1/3	PA, IN, VA, LA, WA, IA, OR, MS, MI, NC, CT, AR (12)	CO, KS, OK, UT (4)
Highest 1/3	VT, WI, MN, ME, NH, RI, DE, WV (8)	NM, NE, NV, ID, MT, SD, ND, WY (8)
	Chi-square statistic 5.64 Probability of a greater chi-square statistic .059	

Chi-Square Analysis

After categorizing the states with respect to climate, chi-square analysis allows us to examine the federally assisted water-related grant variables. By dichotomizing the data set with respect to arid versus humid states, we are able to determine if a statistically significant difference could be detected within any of the three grouped variables. Economic theory predicts that if something becomes more desirable and scarce then it becomes more valuable, and as its value goes up, society would be more prone to protect it. Water would seem to be a perfect example of this economic theory and one that could be subjected to statistical testing. For the purposes of this study, we are assuming that when states participate in one of these federally assisted water-related grant programs, it is a proxy that society, or the state, is attempting to protect or preserve the water in that state. Therefore, we hypothesize that the scarcity of water associated with arid-semiarid states would lead these states to participate at a much higher level in the federally assisted water-related grant programs. In other words, after adjusting the data for population influences, arid-semiarid states would participate in water grants at a different level than their humid-semihumid counterparts.

Table 7.3.—Summary of Water-Related Programs

Participation Rate	Humid-Semihumid	Arid-Semiarid
Lowest 1/3	VA, FL, MD, IN, IA, IL, WI (7)	CO, NV, KS, UT, NE, OK, AZ, ID, TX (9)
Middle 1/3	GA, NH, AL, WA, MN, NJ, MO, TN, PA, NC, MI, SC, OH, CA, DE, OR (16)	(0)
Highest 1/3	AR, KY, MA, CT, MS, RI, ME, VT, WV, LA, NY (11)	SD, MT, WY, ND, NM (5)
	Chi-square statistic 12.3 Probability of a greater chi-square statistic .0002	

The results of the chi-square analysis for one of the federally assisted water-related grant programs is presented in Table 7.1. Specifically, the results from the variable Clean Water State Revolving Fund that were adjusted by total population of each state can be examined. If there were no difference between propensity for the humid-semihumid states and the arid-semiarid states to participate in this particular revolving fund, we would expect an equal number of states to be in each column. That is, we would expect the number "11" to appear three times under heading Humid-Semihumid States, and the number "5" to appear likewise in the Arid-Semiarid column. (Actually the expected values are 11.3 and 4.66, but 11 and 5 are better given the discrete nature of the data.) This distribution of states in the table indicates that some factor seems to be operating to differentiate the two groups of states. That is, based on the chi-square value of 7.46 and the accompanying level of significance of .024, we concluded that the arid-semiarid states participate in the Federal Clean Water State Revolving Fund at a different rate than do the humid-semihumid states. In fact, looking at the distribution of arid-semiarid states, it appears that these states have a tendency to either fully participate or barely participate in this clean water program, whereas the humid-semihumid states tend to participate moderately in this program. The second federally assisted water-related grant program we tested was the Drinking Water State Revolving Fund program. (See Table 7.2 that shows the level of significance as .059, and a chi-square statistic of 5.64, which we interpret as still being statistically significant for this study.) Again, we concluded that there is a difference between the arid-semiarid and the humid-semihumid states' participation in the Drinking Water State Revolving Fund. A comparison of the two columns of Table 7.2 reveals a distinct pattern of increasing participation by arid-semiarid states and decreasing participation by humid-semihumid states as funds are disbursed on the basis of population. This pattern highlights the clear difference between the individual water needs of these two groups of states and validates the economic theory of scarcity.

The final federally assisted water-related grant program we subjected to a chi-square analysis was a Summary of Water Related Programs, again weighted by population (see Table 7.3). Notice that the chi-square statistic has a value of 12.3 with a level of significance of .0002 that clearly indicates the difference in level of participation of the arid-semiarid states compared to the humid-semihumid states. Again, it is interesting to observe that arid-semiarid states do not participate moderately in federally assisted water-related grant programs when viewed in the aggregate. This is an interesting result because the arid-semiarid states seem to migrate toward little or great participation in these programs.

THE GREAT PLAINS

Defining the geographic boundaries of the Great Plains is not as easy a task as it would seem. The broadest delineation (and the one used in this chapter) defines the Great Plains from the Texas-Mexico border up through the central states (Texas, New Mexico, Oklahoma, Kansas, Colorado, Nebraska, Wyoming, South Dakota, North Dakota, and Montana) into Canada and includes the vast stretches of Manitoba, Saskatchewan, and Alberta. It is interesting to note the influence that the lack or scarcity of water has upon this definition because all Great Plains states are classified as arid-semiarid. Further reinforcing the importance of weather and water upon the delineation is the fact that the Ogallala Aquifer lies under all included states except North Dakota and Montana. Obviously the analysis of arid-semiarid states versus humid-semihumid states is of particular importance to the Great Plains states. Of all states classified as arid-semiarid for analysis purposes, only four states are not included in the Great Plains: Arizona, Nevada, Utah, and Idaho.

A review of the analysis and conclusions reveal that the Great Plains states are totally representative of the arid-semiarid states in general. As seen in Table 7.1, the states of Colorado, Oklahoma, Texas, New Mexico, Kansas, and Nebraska ranked at the bottom for participation. The states of South Dakota, North Dakota, Montana, and Wyoming ranked in the upper third of all states for grant participation. In other words, the Great Plains states either participated or they didn't. There was no middle ground. The Drinking Water State Revolving Fund (see Table 7.2) reflects more diverse grant participation from Texas (low participation) to Colorado, Kansas, and Oklahoma (middle participation) to New Mexico, Nebraska, Montana, South Dakota, North Dakota, and Wyoming at the highest participation level.

A summary of grant participation in all six water-related EPA programs for the ten Great Plains states is revealed in Table 7.3. The Great Plains states of Colorado, Kansas, Nebraska, Oklahoma, and Texas were all classified as being in the lowest participation category, whereas South Dakota, Montana, Wyoming, North Dakota, and New Mexico were in the highest participation category.

It is obvious once more that the participation of Great Plains states in water-related grant programs is statistically significant. The Great Plains states either participate or do not participate. They do not migrate toward the middle. Chi-square analysis of the ten Great Plains states would have an expectation of three and one-third states in each of the grant participation categories (lowest, middle, and highest). In the summary of the six EPA Water Related Programs, six states were categorized as being in the lowest participation and four states in the highest participation category. There were no Great Plains states in the middle category. This behavior is what accounts for the analysis being statistically significant for the Great Plains states in particular and arid-semiarid states in general.

REFERENCES

Britton, C. R., R. K. Ford, and J. G. Hehr. 1993. Commercial Banking Conditions in the Western United States: An Arid/Non-Arid Comparison. *Forum of the Association for Arid Lands Studies* 8, no. 1: 97–106.

Congressional Budget Office. 1997. *Federalism and Environmental Protection: Case Studies for Drinking Water and Ground-Level Ozone*. ACBO Study. Washington, DC: Congress of the United States.

Merrill Lynch. 1998. *Theme and Profile Investing: Special Report*. Vol. 21. New York: Author.

Oates, Wallace E. 1998. Thinking About Environmental Federalism. *Resources*. (Winter): 14–16.

Eight Water for the Future: The Development of Markets in the Texas Plains

David W. Yoskowitz

Many communities across the western United States, and in particular the Great Plains, are beginning to feel the pressure of dwindling water supplies, especially those that rely on groundwater. The numerous droughts throughout the 1990s have made this pressure all too clear. Alternatives become difficult to find when there is very little if any additional water to be captured. One alternative that is slowly developing in the Plains is that of water marketing. The concept is simple: water is traded like any other commodity given the laws that apply. The result is water moving to its higher-valued use.

As cities in the West Texas Plains continue to grow, so does demand on the limited resource—water. In a region where the average rainfall ranges from 16 to 20 inches in good years, the supply of this resource is of constant concern to those individuals responsible for delivery to their customers. The practice of water marketing helps to alleviate some of the pressure that develops from both demand and supply sides.

A market for water exists because of its scarcity. In the state of Texas, as well as the majority of other western and southwestern states, most of the water has been claimed. There is very little new water to be had. Therefore, the only options for individuals, cities, and industry in search of new water supplies is to buy water from other right holders. Very active markets have begun to develop in regions that are feeling the pressure of dwindling supplies. Along the Rio Grande in Texas, a spot market for water has seen over a thousand transactions in five and a half years (Yoskowitz, 1999). The California Drought Water Bank was established in 1991 to temporarily facilitate the transfer of water from right holders in Northern California to thirsty cities in Southern California and proved quite successful. This new approach to garnering water supplies has now begun to find its way into the West Texas Plains.

The cities of Midland, Lubbock, Plainview, and Amarillo represent the four largest municipalities in this region. It is some of that experience that is presented in this chapter.

HISTORY AND GROUNDWATER LAW

Although there has been active purchasing of groundwater in the West Texas Plains since the middle of the twentieth century, there has been very little work examining it. Jonish and Butler (1983) noted that certain member cities of the Canadian River Municipal Water Authority (CRMWA), which comprises several municipalities on the West Texas Plains, are actively searching for groundwater rights to make up for projected shortfalls. The CRMWA provides surface water to its members cities via an extensive pipeline system, but in times of high demand, shortfalls may exist for some municipalities. The purchase of groundwater helps alleviate the pressure.

Griffin and Boadu (1992) explain that groundwater marketing in Texas is limited to two forms. The landowner may "reduce groundwater to ownership" by pumping the water that can then be sold and transported. Or the land itself may be sold, after which the new owner may pump water and apply it to a different use, perhaps somewhere else.

There are four possible market failures associated with groundwater marketing.

1. Return flow externalities, where return flow to the aquifer and groundwater creates third-party effects that are relevant to market exchanges.
2. Secondary economic effects also at issue for groundwater exchanges—again giving rise to the difficult question of whether to protect sectors that are economically linked to water right sellers.
3. Intertemporal externalities, which are concerned with the decisions of today's users on future users of groundwater. If profit-maximizing individuals are only concerned with their own wellbeing and not that of future generations, then excessive use of nonrenewable groundwater stocks will occur, not allowing those future generations opportunities for consumption.
4. Drawdown externalities, if groundwater rights are not carefully defined and these external effects between users go uncompensated.

As surface water rights become scarce, municipalities in the West Texas Plains are turning to groundwater resources to supplement current supplies for future use. Therefore, an understanding of Texas groundwater law is important in the analysis of water marketing in this region.

In contrast to surface water law, which was crafted by the Texas Legislature, Texas groundwater law is judge-made law, derived from the English common law rule of "absolute ownership." Groundwater belongs to the owner of the land above it and may be used or sold as private property (Kaiser, 1993). However, Griffin and Boadu (1992) note that it has been argued that the "absolute ownership" doctrine is a misnomer for Texas groundwater principles, because the groundwater user is not protected from water table declines brought on by one's neighbor. The practical effect of Texas groundwater law is that one landowner can dry up an adjoining landowner's well and the landowner with the dry well is without a legal remedy. However, there are five situations in which Texas landowners can take legal action for interference with their groundwater rights:

1. If an adjoining neighbor trespasses on the land to remove water either by drilling a well directly on the landowner's property or by drilling a "slant" well on the adjoining property.
2. There is malicious or wanton conduct in pumping water for the sole purpose of injuring an adjoining landowner.
3. Landowners waste artesian well water by allowing it to run off their land or to percolate back into the water table.
4. If there is contamination in a landowner's well.
5. Land subsidence and surface injury result from negligent over-pumping from adjoining lands (Hutchins, 1961; Kaiser, 1993).

In addition to these five situations, there is land that falls under the laws and regulations of an underground water district. In 1949 and again in 1985, the Texas Legislature passed a law authorizing, only with local voter approval, the creation of these districts. The districts by Sections 52.001 and 52.501 of the Texas Water Code have as their purpose the prevention of waste, the prevention of land subsidence, protection of water quality, and conservation of groundwater supplies. All underground water districts tread a narrow path between private ownership rights of groundwater and responsibility to protect the water resource (Kaiser, 1993).

The trend in the West Texas Plains has been to purchase land, water rights, or both for municipal use. The Texas Supreme Court in *Corpus Christi v. Pleasanton*, 154 Tex. 289, 276 S.W.2d 798 (1955), held that under common law, "There certainly was no limitation that prohibited the use of the water off the premises where it was captured. Neither was there any restriction of its use to a particular area" (Hutchins, 1961, n.p.). This aspect of groundwater law allows the municipalities to engage in transactions that will transport groundwater off-site to be used by the municipalities. In addition, it is not necessary that the municipalities purchase the water right; they may just lease it.

MUNICIPAL WATER TRANSACTIONS

Data

Water utilities personnel at each of the four cities were asked to respond to a structured questionnaire about each groundwater transaction that their municipality was involved with. Categories of information included:

- date of purchase
- type of purchase (i.e., groundwater rights, real property, etc.)
- number of acres purchased
- price per acre
- location of purchase
- any options on the purchase
- if the purchase is currently being used or held as inventory
- what has been or what will be the cost of development
- any legal complications
- other options considered

Prices were adjusted to 1984 levels using the Consumer Price Index. Nominal prices are also given. The price per acre-foot of recoverable water was provided as another means of comparing and contrasting purchases. The cities of Amarillo and Midland provided estimates of recoverable water. For Lubbock, the price was determined by the number of acres the purchase covered multiplied by the saturated thickness at the time of purchase. Total saturated area was then multiplied by .15, representing the estimated specific yield given the geological characteristics of the region. This figure was then divided into the purchase price (1984 dollars). Information on saturated thickness was obtained from Geraghty and Miller, Inc. (1992) and the Hydrological Atlas for the High Plains Underground Water Conservation District No. 1 (1990). Analysis of the data will be used to reveal any patterns within a municipality and across municipalities (see Table 8.1).

Midland

The City of Midland has made only one purchase of groundwater to augment their supplies. This purchase was made in 1966 in Winkler and Loving Counties, west of the City, that consisted of 20,650 acres of real property along with the water rights for $2.31 million (1984 dollars). The City estimated that there is 669,625 acre-feet of recoverable water at a price per acre-foot of $3.45. Currently the water is being held as inventory until the year 2030, when the city expects that they will need the water. Five to ten years before then, construction will begin to develop the well field at an estimated cost of $48.6 million (City of Midland, personal interview with water utilities personnel, December, 1996).

The City has also recently entered into a contract with the Colorado River Municipal Water District (CRMWD) to be supplied water under a "take-or-pay" contract. The city is obliged to take the water at nearly a constant rate and make up shortfalls with their current groundwater supplies from the Paul Davis well field north of the city. This well field is leased from the University of Texas on a fifty-year contract (HDR Engineering, 1994).

The volume of water the City must take increases each year through the year 2030. The City is not a member of the CRMWD, just a purchaser of water, and is therefore required to pay $.032 per thousand gallons over what their neighbor, the City of Odessa, which is a member city, pays (City of Midland, personal interview, 1996).

With their eye on current water supplies in use and inventoried, Midland is still focusing on the long-term problems of supply. A few of their options include purchasing groundwater rights in Gaines County and the foothills of the Davis Mountains. They have also conducted a feasibility study on the implementation of a reverse osmosis system that would provide the opportunity to use sources of water previously unavailable.

Lubbock

The City of Lubbock is most active in purchasing new sources of water. The Ogallala Aquifer is much thicker north and west of Lubbock than it is south and east of Lubbock. North of Plainview (which is north of Lubbock) there is more groundwater of better quality to be captured than south of Plainview. Cities in the northern sections of the Panhandle have also had an easier time drilling for water on existing property, whereas Lubbock has consistently had to look elsewhere.

The City of Lubbock (population of approximately 200,000) has been very active in purchasing land and groundwater rights to bolster their supplies, as Table 8.1 indicates. The City has been drilling on its own property since the early 1900s. However, due to the growth of Lubbock, administrators decided in the 1950s to begin looking outside the city limits for additional supplies of water.

The first purchase of water rights took place in January of 1953 in Bailey County, northeast of the City. A mediator purchased 57,000 acres of land with the understanding that he would then sell the water rights to Lubbock. The purchase price to the rights of that water (in 1984 dollars) was $37 per acre ($10 nominal value). The price per acre-foot of recoverable water is $.02. Since this first purchase, the City has been very active: it has either actively pursued the purchase of water rights or, as is usually the case, been very amenable to individuals who come to the City willing to sell their rights. Currently, water rights are being purchased for around $300 nominal value an acre (City of Lubbock, personal interview with water utilities personnel, February, 1997). Lubbock, as a rule, conducted a feasibility study and then decided to purchase or not based upon the results of the study.

Table 8.1.—Municipal Water Purchases

Transaction	Date	Type of Purchase	Size of Purchase	Price*	Options or Inventoried	In Use	Development Cost
Midland #1	1966	Land and water rights	20,650 acres	$112 per acre ($36) *$3.45 af*	None	Inventory	$48.6 million
Lubbock #1	1953	Groundwater rights	57,000 acres	$37 per acre ($10) *$.02 af*	Several wells reserved by owner	In use	n/a
Lubbock #2	1953	Groundwater rights	249.4 acres	$374 per acre ($100) *$22.6 af*	Owner maintains right for normal domestic use	Inventory	n/a
Lubbock #3	1957	Groundwater rights	16,047.7 acres	$124 per acre ($35) *$3.30 af*	Owner maintains right for normal domestic use	In use	n/a
Lubbock #4	1975	Real property sold. Own water rights	531 acres	$418 per acre ($225) *$12.96 af*		In use	n/a

	Year		Acres	Price		Status	
Lubbock #5	1985	Groundwater rights	1289.6 acres	$278 per acre ($300) *$11.88 af*	Owner maintains right for normal domestic use	Inventory	$500,000 per 640-acre section
Lubbock #6	1987	Groundwater rights	1464 acres	$264 per acre ($300) *$11.35 af*	Owner maintains right for normal domestic use	Inventory	$500,000 per 640-acre section
Lubbock #7	1988	Groundwater rights	2163 acres	$244 per acre ($289) *$9.19 af*	Owner maintains right for normal domestic use	In use	Minimal, infrastructure in place
Amarillo #1	1984	Groundwater rights	960 acres	$250 per acre *$16.23 af*	Inventory		n/a
Amarillo #2	1988	Groundwater rights	25,459 acres	$99 per acre ($117) *$9.01 af*	Inventory		n/a

*All prices are adjusted to the 1984 price level. Nominal prices are in parentheses. Prices in italics are the price per acre-foot of recoverable water.

The feasibility studies include information regarding the amount of water that the purchase will produce and the capital costs associated with that purchase. The capital costs along with the electrical costs are broken down per thousand gallons to provide a total cost per thousand gallons. The study also included a proposed pumping plan for the site that covers the number of days per year that the site will be pumped and the pumping life of the purchase.

The City has the majority of its water rights in Bailey County. Since the first purchase in 1953 the City has been active in trying to purchase rights that are adjacent to the original purchase (Lubbock #1) or reasonably close. The reason for this is that the infrastructure is in place to transport this water to the City. Therefore, additional property that is located near this infrastructure will keep the cost of developing the well fields down. Beginning in 1975, the City purchased groundwater rights in Bailey County at a price of $418 (1984 dollars). Two of these purchases, Lubbock #5 and #6, are currently being held as inventory with the price per acre-foot of recoverable water at $11.88 and $11.35 per acre-foot, respectively. The cost of development for these purchases is projected to be around $500,000 per 640-acre section. The last purchase the City made (Lubbock #7) was 2,163 acres in 1988 and it is currently in use. Price per acre-foot of water is $9.19 and the cost of development was minimal because the majority of infrastructure was already in place (City of Lubbock, personal interview with water utilities personnel, February, 1997).

In 1953, the City also purchased 249.4 acres of groundwater rights in Lubbock and Hockley County. The price of this purchase was $374 per acre (1984 dollars) and $22.60 per acre-feet of recoverable water, which was well above the purchase price of the rights in Bailey County of the same year (City of Lubbock, personal interview with water utilities personnel, February, 1997). No explanation for the price differentials was available.

One characteristic of the majority of Lubbock's transactions has been the right of the real property owners to normal domestic use of the groundwater. This means that the landowner has a right to use groundwater, usually for domestic purposes that may include household needs and small amounts of irrigation for landscape, gardening, and livestock. In addition, there has been no recorded legal complications with any of the groundwater purchases made by the City.

Lubbock is a member of the CRMWA and fills 80 percent of its water demand from this source, with the remainder made up by the City's ground water supplies. The Bailey purchases make up the largest part of Lubbock's groundwater supply; the two hundred wells need to be producing over one billion gallons a year to meet current demand with the majority of that coming in the summer.

Plainview

The City of Plainview has not been involved in any purchases of underground water rights to supplement current supplies. All of the City's needs are met by the fifteen wells located within the city limits (50 percent) and their allocation from the CRMWA (50 percent). Obtaining any more water will be as simple as drilling additional wells on City property. The City has recently incorporated more land that makes it eligible for wells to be drilled on if necessary. Although not a direct purchase, the incorporated land can be thought of as adding to the groundwater rights of the City (City of Plainview, telephone interview with water utilities personnel, March,1997).

Amarillo

The City of Amarillo is fortunate to find itself in a region of the West Texas Plains overlying a section of the Ogallala Aquifer that is thick with high-quality water. The City has enjoyed a large and easily accessible supply of water. In addition to the groundwater supply, the City is a member of the CRMWA where it has consistently used its full allocation of lake water.

The City has a number of groundwater rights on land that is not within the City limits. However, a number of these purchases were made early in the life of the City and records are not accessible, according to water utilities personnel (City of Amarillo, telephone interview with water utilities personnel, April, 1997). Two recent transactions have been recorded. In 1984, Amarillo purchased 960 acres of groundwater rights in Carson County. The price was $250 per acre for a total of $240, 000 with the price per acre-foot of recoverable water at $16.23. The water is currently being held as inventory (City of Amarillo, telephone interview, April, 1997).

The most recent purchase took place in 1988 and involved 25,459 acres located in northeast Potter County. Groundwater rights only were involved with the real property rights remaining with the land-owner. The price of $99 per acre (1984 dollars) generated a total nominal price of $2,999,833.97. Price per acre-foot is $9.09. The water is currently being held as inventory and there is no projection of when it will be developed (City of Amarillo, telephone interview, April, 1997).

CONCLUSIONS

The groundwater transactions of four municipalities—Midland, Lubbock, Plainview, and Amarillo—along with the Canadian River Municipal Water Authority have been examined. Use of the Case Cluster method has allowed each transaction that has taken place to be a unit of analysis in the case study. This method produces results that can easily be analyzed for patterns within municipalities and across municipalities.

The majority of purchases (six) have been made within the last twenty-two years and a majority of those (five) took place in the 1980s. This increase in purchases is most likely the result of dwindling supplies and increases in demand. Plainview is the only city with no new purchases and a flat growth rate in population, whereas the other municipalities have experienced small to moderate growth rates in population. In addition, the majority of purchases (six) are being inventoried for use at a later date when current supplies are insufficient. These purchases may be inventoried even longer because the three cities are members of the CRMWA (Amarillo, Plainview, and Lubbock). The City of Midland has also recently entered into a contract to receive surface water from the Colorado River Municipal River District that should keep their sole purchase in inventory even longer.

Jensen (1987) observed that when the land is purchased, many cities lease it back to agricultural producers who can use the land in dryland production. This case study found only two purchases in which the real property was bought along with the water rights. In one of these transactions (Lubbock #4), the real property was

135

eventually sold. The municipalities seem to prefer only dealing with the water rights.

Three different prices were examined in relation to the purchase of groundwater. The prices collected from the municipalities were the nominal values per acre. It should be noted that, where the real property was not purchased, water rights were still sold per acre of land. Nominal prices have ranged from $10 per acre in 1953 to $300 in 1987 across municipalities. For the three purchases that the City of Lubbock made in the 1980s, the nominal price has been in the narrow range of $289 to $300 per acre. The recent purchases made in Roberts County by the CRMWA were priced at $339 per acre.

Adjusting the nominal prices for inflation allows the purchases to be compared on a same-year basis. Prices were adjusted to 1984 using the Consumer Price Index. Once again prices range from $37 to $418 per acre across municipalities. However, in the 1980s prices began to settle in the middle $200 range for both Lubbock and Amarillo (except for Amarillo #2). Because the purpose of the purchases is to mine water, prices per acre-foot of recoverable water is also of interest. Beginning in 1975, the price per acre-foot of recoverable water began to fall into a range of between $9 to $13 (except Amarillo #1). Before 1975 price per acre-foot ranged from $.02 to $22.6. The narrow range of prices beginning after 1975 may be an outcome of an increase in technological know-how and information. However, price differences still exists and could be a result of quality, location, and economies of scale.

The buying and selling of groundwater on the West Texas Plains will no doubt continue to be the manner in which water-strapped municipalities find new sources, with surface water limited and fully appropriated. It will also become important for cities to make better use of their wastewater, possibly even filtering it to a quality where it once again is potable.

Needed cooperation between political regions, whether they are cities, counties, states, or countries, is already a reality in managing our joint resources. Doubling these efforts in the case of water will be necessary, as life cannot exist without it. For the United States, this means exploring the possibilities of creating a market for water with both Canada and Mexico in such a manner that all three countries can better manage their water resources.

REFERENCES

Geraghty and Miller, Inc. 1992. *Comprehensive Groundwater Management Study for the City of Lubbock.* Midland, Texas, April.

Griffin, R., and F. Boadu. 1992. Water Marketing in Texas: Opportunities for Reform. *Natural Resources Journal* 20, no. 2 (Spring): 262–88.

HDR Engineering. 1994. *Potable Water Demineralization Feasibility Study.* Dallas, Texas. January.

High Plains Underground Water Conservation District No. 1. 1990. *Hydrological Atlas for Bailey, Lubbock and Hockley Counties, Texas.*

Hutchins, W. A., 1961. *The Texas Law of Water Rights.* The State of Texas, Board of Water Engineers.

Jensen, Ric. 1987. The Texas Water Market. *Texas Water Resources* (Spring): 3-4

Jonish, J., and C. Butler. 1993. Municipal Water Pricing Practices in West Texas. *Texas Business Review* (March/April): 88–94.

Kaiser, R. 1993. *Handbook of Texas Water Law: Problems and Needs.* College Station: Texas Water Resources Institute, Texas Agricultural Experiment Station, Texas A&M University.

Yoskowitz, David. 1999. Spot Market for Water Along the Texas Rio Grande: Opportunities for Water Management. *Natural Resources Journal* 39, no. 2 (Spring): 345–55.

Part Three

Beyond the Plains:
Additional
Considerations

Thirst

They shoveled the dry bee hives as the hot wind weaned into each porthole of the earth as they swept the charred wings like a good penny into the bottom of the wishing well once a gathering place for the children at the lilac spring. They dug deeper for water between the desert ash and blackened skulls of the animals far below their own long shadows as the white heat above their heads evaporated the dew in their eyes. *Even a mirage was a miracle.* The fire spread over the trees into the plague houses and into the hollow ground. Had they broken the covenant of the soil and the bread of the water demons in their last breath and finally understood the language of water?

Charles Fort

Land, Water, and the Right to Remain Indian: The All Indian Pueblo Council and Indian Water Rights

Steven L. Danver

If you want to learn something, take twenty years to study the Indians.
—Martin Vigil, Tesuque Pueblo, 1970

By the first decades of the twentieth century, irrigation had gradually changed the face of Native America. Many tribes, through reservation life or severalty, were being forced to take up a life of sedentary farming. Whether tribes had to adapt from a shifting subsistence economy to intensive agriculture, like many of the tribes on the Plains, or whether they were already an established agricultural society such as the Pueblos of New Mexico, farming on marginal lands became a vital part of reservation life. Irrigation technology allowed the practice of one-crop agriculture to spread from the more fertile Eastern Plains, to the less productive High Plains, to the deserts of the Southwest. The 1908 Supreme Court decision in the case of *Winters v. United States* (207 US 565, 573, 575-7), which dealt with a water dispute between Indians and whites in Montana, theoretically established preeminent water rights for Indian tribes. After the *Winters* decision appeared to confirm Indian water holdings in the long term, the Bureau of Indian Affairs (BIA) began building elaborate irrigation systems. However, a lack of maintenance by the BIA dashed the government's hopes of Indian self-sufficiency that irrigation appeared to promise (Lewis, 1995). The ineffectiveness of early efforts to develop Indian water did not deter whites from irrigating to the extreme, as can be seen in the fact that by the 1980s, the lower water tables in Texas, California, Nebraska, and Kansas had been seriously depleted (Fixico, 1998).

Although Indians of the Plains and the Southwest have had issues of land and water in common, it has long been asserted that the Plains tribes were less reliant on agriculture than were the Pueblos (Lowie, 1954). Accordingly, simple logic would dictate that in a semiarid environment such as New Mexico, scarce water and arable land would necessarily become precious resources. Thus, in the multiple interactions between the Pueblo peoples of New Mexico and the Anglo and Hispanic populations of the region, there are not many issues more important than land and water rights. Accordingly, there are no more vital topics when it comes to the relationships between the Pueblo nations as sovereign political entities and the local, state, and federal government agencies that control policy in these important areas. The relative importance of these issues can be deduced from the most cursory studies of Pueblo life during the periods of Spanish and Mexican government, but most importantly, the Pueblos themselves acknowledge this fact. The loss of water rights over the twentieth century has been asserted to have constituted a direct threat to the Pueblos as "viable, permanent and independent communities" (Rosenfelt, 1969, n.p.). Correspondingly, tribal sovereignty is both important and necessary for the Pueblos to exercise any influence over government policy in these vital issues.

In "Reflections in a Ditch," the introduction to his influential book *Rivers of Empire,* Donald Worster (1985) asserts that

> [t]here is nothing harmonious, nothing picturesque about the western world that has developed beside the irrigation ditch. There is little peace or tidiness or care, little sense of a rooted community. There is no equitable sharing of prosperity. The human presence here often seems very much like the tumbleweeds that have been caught in the barbed-wire fences: impermanent, drifting, snagged for a while, drifting again, without grace or character, liable to blow away with a blast of hot desert wind. (6)

This brief statement presents a picture of the American West that has become a part of the "new Western history." This view of history has gone a long way toward correcting the ways historians have traditionally looked at the West, and yet there is something missing here. Worster's picture may reflect white settlements and attempts at the large-scale change and irrigation of the West, but it would be inaccurate to place this same evaluation on the long-term irrigation of the American West that has been a part of Pueblo and other

American Indian cultures for millennia. Of course, Worster discusses the Pueblo and other American Indian activities as a part of what he calls the "local subsistence mode" of irrigation, but one could hardly say that tribal groups like the Pueblo nations have been "impermanent, drifting, snagged for a while, drifting again, without grace or character" (1985, 6). Indeed, their very permanence has depended on the scarce waters of one of the driest regions of the arid West.

However, there is much that American Indian historians can draw from Worster's seminal work. More important, he has contributed to the field of the history of the American West by pointing out the equation of control of Western waters with social and political power in the West, and this is certainly the case when it pertains to the Indian nations of the West. Debate and litigation continue between tribal groups, such as the Pueblo nations, and state and federal governments over the rights to the water and the land that are so important in shaping the character of the American West. Of course the power has usually been in the hands of decision makers within the federal government who have historically proven impervious to Indian demands. The story of the Pueblo nations in the twentieth century has been one of attempting to use what power the nations could muster in order to protect rights to land and water that are central to their identities and relationships with the world.

During the colonial periods of New Mexican history, both Spanish and Mexican authorities recognized that their citizens would come into contact and conflict with the Pueblos, and had specific laws dealing with how the inevitable disputes over the land and water rights of the Pueblos would be adjudicated (Weber, 1982, 1992). Hence, land and water law and interracial relations in New Mexico had a long-established history when the government of the United States took over the entire Southwest (Mexico's northern provinces of Alta California and Nuevo Mexico) at the end of the Mexican War. But when the United States took over the region with the Treaty of Guadalupe Hidalgo in 1848, they also took over responsibility for the protection of these established rights of the Pueblos and other tribes.

This chapter primarily examines the land and water rights disputes between the Pueblo Nations and white Americans over the course of the twentieth century. This statement of purpose expresses nothing novel, nothing that has not been attempted and done well in numerous articles and books. And yet most of the articles and books written on Indian water rights in general and Pueblo Indian water rights in particular share one common characteristic—most of these studies proceed from a "top-down" perspective, mostly examining the federal and state legal actions that govern water rights. They have tended to pay very little attention to the Indian actions in response to and often in defiance of these governmental actions. This is not to downplay the importance of understanding the political aspects of this history. Truly, an understanding of the various complex and often conflicting government actions in relation to land and water rights is necessary to a proper understanding of the situation in the Pueblos, and an examination of the prior works on these subjects will serve this study well as background material.

However, as Angela Cavender Wilson (Dakota) lamented both at the 1998 Western History Association meeting and in her chapter of a recently published book on writing American Indian history, historians often claim to portray an "Indian perspective" in their work, and yet they seldom do include Indian voices. She went on to state that the field has been and to an extent still is "dominated by white, male historians who rarely ask or care what the Indians they study have to say about their work" (Wilson, 1998, 23). What will distinguish this history of Pueblo Indian land and water rights from what has come before is that this chapter will come from an approach that examines the various proactive and reactive operations that the Pueblo people, the Pueblo Nations, and their cooperative body, the All Indian Pueblo Council, have taken in defense of their land and water holdings. This chapter will include the legal aspects of land and water adjudication in New Mexico, but will emphasize the voices of Pueblo people who have been involved in various capacities in the process. Only by understanding the governmental and tribal actions can we truly comprehend the vitality and long-lasting importance of water and land issues to all of the populations in New Mexico. The first step in the process of examining the Pueblo Nations' own role in this process is to investigate the origins of intra-Pueblo cooperation through the All Indian Pueblo Council. Then the various incidents, legislation, and court cases effecting Pueblo water rights are examined in the context of the Pueblo nations' own actions taken to protect those rights.

AMERICAN INDIAN LAND RIGHTS: AN INTRODUCTION TO THE EVOLUTION OF FEDERAL POLICY

> The United States Government has made many pious utterances about the protection of tribal land and water interests.
> —All Indian Pueblo Council, 1973

Before delving into the Pueblos' role in securing their land and water rights, it would be helpful to give a brief overview of the various judicial and legislative actions that form the background for discussing Indian land and water rights within any tribal context. Whereas the decisions affecting American Indian water rights are complex and often contradictory, Indian land law is a much more simple, although no less contentious issue. Basic American Indian land law was established very early in American history, through the Supreme Court decisions of Chief Justice John Marshall. The decisions of the 1830s affecting the Cherokee and other Southeastern tribes laid the foundation of the idea of tribal sovereignty over land. In the *Cherokee Nation v. Georgia* decision (30 US, 5 Pet., 1, 1831), Marshall declared that the tribe constituted a "domestic dependent nation" whose members were to be considered wards of the United States, and whose rights were to be protected by the federal government (Deloria and Lytle, 1983, 1984).

The next year, in *Worcester v. Georgia* (31 US, 6 Pet., 534-36, 558-63, 1832), Marshall explained his idea of tribal sovereignty in greater detail when he declared that "Indian nations had always been considered as distinct, independent political communities, retaining their original natural rights, as the *undisputed possessors of the soil* [emphasis added]." Although President Andrew Jackson saw fit to ignore the Supreme Court's decision confirming the Cherokee land rights and moved them across the Trail of Tears to Indian Territory in Oklahoma, Marshall's opinions in these cases have been cited throughout American judicial history in a multitude of cases as setting the precedent for confirming tribal land rights. However, that precedent was severely tested later on in the century through various federal actions aimed at forcing assimilation on American

145

Indians. Federal policy as set out in the General Allotment Act of 1887 (more popularly known as the Dawes Severalty Act) attempted to do away with the tribes' land base through the distribution of 160-acre parcels of reservation land to individual American Indians. Not only did this act proceed with the breakup of tribal reservation land, challenging the widely held American Indian idea of communal land ownership, but it also made those Indians who received their allotments United States citizens, making them subject to the laws and taxes of the state.

Even the Supreme Court confirmed this change in American Indian land policy in deciding the case of *Lone Wolf v. Hitchcock* (181 US 553, 564-68) in 1903. When a Kiowa man named Lone Wolf protested the severalty of the Kiowa-Comanche Reservation by claiming that it went against the provision in the 1867 Treaty of Medicine Lodge that stated that no part of the reservation could be sold or ceded without the approval of three-fourths of the adult male tribal members, the court seriously damaged tribal land rights by declaring that "[p]lenary authority over the tribal relations of the Indians has been exercised by Congress from the beginning, and the power has always been deemed a political one, not subject to be controlled by the judicial department of the government." In other words, the court stated that it was no longer the final recourse for Indian tribes attempting to protect their land rights through sovereignty against intrusion by the federal government. The Congress alone was responsible for American Indian policy, and their decisions were not open to judicial review.

The situation for the Pueblo nations was even more desperate during 1900-20. Because their land holdings had been conferred by the Treaty of Guadalupe Hidalgo and not by act of Congress, the Pueblos were considered to be excluded from the provisions that set out federal relationships with American Indians as outlined in the Trade and Intercourse Act of 1834. This status granted the Pueblos citizenship as landowners, and therefore opened them up to taxation and subjugation to the laws of the state of New Mexico. Regarding land law, this made the Pueblo nations subject to the same statutes that governed all land holdings in the state. Any recourse taken to dispute whites' squatting on Indian lands would have to be taken up with the New Mexico state courts, who had earned a reputation of hostility toward the Pueblos, and not with the federal government or the Bureau of Indian Affairs.

This dim view of Indian land rights and sovereignty in general, and Pueblo rights in particular, would not be overturned until the passage of the Wheeler-Howard Act of 1934 (popularly known as the Indian Reorganization Act), which stated that "hereafter no land of any Indian reservation, created or set apart by treaty or agreement with the Indians, Act of Congress, Executive order, purchase, *or otherwise* [emphasis added], shall be allotted in severalty to any Indian." For the Pueblos, reform came to fruition nearly a decade before the Wheeler-Howard Act was passed, although largely at the impetus of the same man, John Collier. Their exclusion from normal American Indian status had been overturned in the case of *United States v. Sandoval* (231 US 28) in 1913. Through this decision, the Supreme Court placed the Pueblos into the same trust relationship with the federal government as other tribes, reversing the exclusion to the provisions of the Trade and Intercourse Act of 1834 that had governed the Pueblos since their incorporation in the United States. Furthermore, this decision placed the responsibility for protecting the rights of the Pueblos with the federal government. In practical terms, however, this distinction of federal policy did relatively little to improve or protect the land and water rights of the Pueblos. Reforming federal policy toward Pueblo land and water would have to wait until 1924. However, before investigating Pueblo efforts to reform federal policy, water policy as it existed during the period before the 1920s needs to be explained.

AMERICAN INDIAN WATER RIGHTS: AN INTRODUCTION TO CONFUSION

That river has been flowing for centuries and it is doing its duty today. Now we want protection because we depend on its life.
—Daniel Mora, Jemez Pueblo, 1927

If federal Indian land policy was the main point of contention and conflict during the nineteenth century, then the even more confusing and ambiguous area of federal Indian water policy was the main point of contention during the twentieth century. Water policy as it relates to American Indians has been determined by the implementation of two contradictory methods of adjudication: the prior appropriation doctrine, and the reserved rights (or *Winters*) doctrine. Another point of conflict is that, like land policy, there have been constant "turf wars." The state and federal governments at various times have both claimed jurisdiction over Indian water. This particular issue gets even more complicated when one considers that often both Indian and white water users are involved in the adjudication of the same source. A final point of contention is that the federal government has not dealt with Indian water rights through statute law or act of Congress. The federal courts have established the reserved rights doctrine through case law, but case law in this contentious area is always subject to interpretation and change as well as being extremely difficult to implement (Hundley, 1978, 1982; McCool, 1987, 1994).

The federal government's American Indian water policy is theoretically based on federal court decisions, as there have not been many definitive, all-encompassing water rights bills passed by Congress. The most-cited court decision regarding Indian water rights took place on the northern Plains in relation to a dispute between white farmers and the Gros Ventre and Assiniboine tribes on the Fort Belknap Reservation in Montana. This court decision has formed the most generous basis for Indian water rights, and thus the most contention with white water-claimants. The Supreme Court opinion in the case of *Winters v. United States* (207 US 565, 573, 575-77) in 1908 sought to adjudicate the water rights of the Milk River that formed the northern border of the reservation. Both Indians and non-Indians claimed the waters of the Milk, as it was the only reliable source of water in the region. Further, the BIA had promised to develop irrigation for the reservation from the river, while the non-Indian settlers had been promised a federal reclamation project to irrigate their lands from the Milk. The Indians claimed that when they ceded aboriginal lands surrounding the reservation they did not cede the water rights associated with that land. The crux of the Supreme Court's opinion in the case was this:

> The lands ceded, were, it is true, also arid; and some argument may be urged, and is urged, that with their cession there was the cession of the waters, without which they would be valueless, and "civilized communities could not be

established thereon." And this, it is further contended, the Indians knew, and yet made no reservation of the waters. We realize that there is a conflict of implications, but that which makes for the retention of the waters is of greater force than that which makes for their cession. The Indians had command of the lands and the waters—command of all their beneficial use, whether kept for hunting, "and grazing roving herds of stock," or turned to agriculture and the arts of civilization. Did they give up all this? Did they reduce the area of their occupation and give up the waters which made it valuable or adequate? (McCool, 1994, 38)

The efforts of the federal government to defend the water rights of the tribes residing on the Fort Belknap Reservation met with the approval of the Supreme Court when it decided that the rights the Indians had as having had "command of the lands and the waters" before non-Indian occupation overrode the claims of the non-Indian settlers. When the federal government set up the reservations, they also, the theory went, reserved the water rights of those Indians, just as the federal government reserved the water rights to any other federal landholdings.

However, the application of what would come to be known as the *Winters* doctrine came up against the prevailing method of adjudicating water claims in the Western United States. The doctrine of prior appropriation, which governs water rights in the West, can be best summed up in the phrase "first in time, first in right." In other words, the first to make beneficial use of the water has the right to all of the water they originally made use of. Whatever is left after the first claimant's use of the water would be the property of the second claimant, and so on down the line (Wilkinson, 1987). When this doctrine is taken to include Indian tribal use, the courts necessarily enter the picture to adjudicate the amounts that the tribe would have the rights to as determined by their historical use of that water source. As Indian tribes are theoretically not held to state laws in these matters, a conflict arises over which water rights doctrine is applicable to the adjudication of rivers that flow over both Indian and non-Indian lands. The *Winters* doctrine would seem to support the view that Indians have the right to all of the water needed to irrigate their claims, and yet the doctrine of prior appropriation supports the idea that, if the Indians did not historically irrigate their lands, then non-Indian water claimants would be substantiated.

As clear a contradiction as this may seem, the actual situation in practice was both less contradictory and more confusing than the federal decisions would make it appear. Daniel McCool has pointed out that these two contradictory water rights theories created a conflict of interest within the Justice Department. As the Justice Department was to be the legal representative for all federal interests, their official position in favor of prior appropriation in the west was in direct conflict with the reserved rights or *Winters* doctrine that was supposed to determine Indian water rights. The *Winters* doctrine theoretically makes the prior appropriation doctrine irrelevant. In practice, however, federal irrigation and reclamation programs were rarely undertaken in Indian interests, even when they were constructed adjacent to Indian lands. The Reclamation Service, dedicated to the doctrine of prior appropriation, acted accordingly when adjudicating the waters made useful by their construction (McCool, 1994). Donald Worster was quite right when he noted that "neither the courts nor Congress managed to settle the issue," and that "white appropriators had an uneasy but clear edge: they were already in possession" (Worster, 1985, 298). Further Supreme Court decisions over the course of the century would clarify the issue legally, but not practically. The BIA felt that the *Winters* doctrine was too vague to be implemented and, because it was based on case law, was on a weaker legal footing than the statutory law that supported the doctrine of prior appropriation.

Much later, in 1963, the federal courts were asked to clarify federal policy when dealing with the adjudication of waters between state governments and American Indian tribes in the case of *Arizona v. California* (373 US 546). This decision of the Supreme Court went a long way toward determining current federal policy toward Indian water rights. The majority opinion argued that

> [t]he doctrine of equitable apportionment is a method of resolving water disputes between States. It was created by this Court in the exercise of its original jurisdiction over controversies in which States are parties. An Indian Reservation is not a State. And while Congress has sometimes left Indian Reservations considerable power to manage their own affairs, we are not convinced by Arizona's argument that each reservation is so much like a State that its rights to water should be determined by the doctrine of equitable apportionment. Moreover, even were we to treat an Indian Reservation like a State, equitable apportionment would still not control since, under our view,

the Indian claims here are governed by the statutes and Executive Orders creating the reservations.

With this opinion, the Supreme Court confirmed the idea of reserved rights that lies at the heart of the *Winters* doctrine. Furthermore, the court clarified the method to be used for adjudicating water claims where a conflict between prior appropriation and *Winters* rights existed (McCool, 1994). In these cases, the Indian water claims are given a very early priority date, no later than when their reservation was first established. Furthermore, these claims can reach very high levels, as determined by the maximum irrigable acreage held by each Indian. As modern technology has made irrigation of previously arid lands more possible, the amount of water that Indians are able to claim under *Arizona v. California* has expanded accordingly (Wilkinson, 1987).

Of course, all of these decisions have occurred over the course of the history of the Pueblo Nations and have taken time to have effects on the day-to-day lives of the Pueblos. Although these laws can be seen as very liberal in their provisions for the needs of Indian tribes, often land- and water-hungry non-Indian settlers and developers have overlooked the legal rights of Indians. Even though the federal government has the responsibility to protect Indian rights, when Indians have not been vigilant about protecting those rights they are often submerged in the states' rush to meet the needs of their non-Indian populations. Meeting the challenge of protecting the land and water rights set out by federal court decisions has forced tribes like the Pueblos to ignore the various differences and disputes that have existed over the centuries between them and form organizations to take action in defending their rights in the legislative and judicial realms.

THE ORIGINS OF THE ALL INDIAN PUEBLO COUNCIL (1900-1959): REVIVAL OF AN ANCIENT CONFEDERATION OR JOHN COLLIER'S CREATION?

The Indians have been meeting for time immemorial.
—Pablo Abeita, Isleta Pueblo, 1929

Gradually the Pueblo Indians got what they want, improve, of course, they [the BIA] didn't done everything what we asked, but at least they really work and they got busy and help us.
—Martin Vigil, Tesuque Pueblo, 1971

Although the Pueblos had advocated for themselves for centuries in dealing with the governments of Spain, Mexico, and the United States, these dealings took on both a new approach and a new urgency during the early twentieth century. By 1916, the Superintendent of the Pueblo day schools wrote to the Commissioner of Indian Affairs that "there is widespread dissatisfaction among practically every tribe of Pueblo Indians because of trespasses upon their lands, either agricultural or grazing, by outsiders, and there is no action being taken to protect their interests" (Superintendent, letter to Commissioner of Indian Affairs, SLTC, March 2, 1916).

In 1921, when a white man fraudulently purchased and built a fence on 65 acres of land belonging to Tesuque Pueblo, a group of local Indians arrived to tear down the fence. Their leader, a twenty-five-year-old named Martin Vigil was singled out by local law enforcement for instigating the disturbance. One of the officers pointed out a government survey marker and asked Vigil, "Can you read?" Vigil answered "No." The lawman proceeded to tell him in Spanish that the marker specified a fine of $200 and a prison sentence for anyone who removed it. Vigil answered, "We don't care about that, this is still Indian land, and nobody asked us if they can put that marker there, or the fence. We know this is our land, so we'll use your court system to prove it" (Sando, 1992, 204).

The conclusions drawn from this story seem obvious, but the implications for the lives of the Pueblo Indians are vast. On May 31, 1921, Senator Holm O. Bursum of New Mexico introduced a bill in

Congress designed to resolve these decades-old land claim disputes arising from the presence of Anglo settlers, or more properly "squatters," on Pueblo land. This first Bursum Bill would have confirmed any non-Indian claims of title to Pueblo land, either with or without color of title, as long as they were held at least ten years prior to 1912. Due to the numbers of such claimants, such a sweeping bill would have virtually destroyed the pueblos' land base, and the Indian Rights Association protested the legislation directly to Secretary of the Interior Albert B. Fall. Fall was able to get Bursum to rework the bill to be, at least theoretically, more fair to the Pueblos. After meeting with both Fall and Commissioner of Indian Affairs Charles H. Burke, Bursum introduced his reworked bill into Congress on July 20, 1922 (Philp, 1981).

Although John Collier, as a social worker who was working under the auspices of the United States Federation of Women's Clubs, was already well aware of the issue, by September the Bursum Bill had caught the attention of numerous national and local groups working for reform in Indian affairs. However, the issue would receive even wider publicity within New Mexico itself when it was given prominent space in the *Santa Fe New Mexican*. The entire text of the legislation was presented in the September 20, 1922 issue, and in a front-page article five days later, John Collier's criticisms of the bill appeared. In this first article on the bill, Collier outlined the objectionable parts of the bill, especially Section 8. In this section, the bill states:

> That all persons who, or corporations which, for more than ten years prior to June 20, 1910, either in person or through their respective predecessors in claim of interest, grantors, privies, or agents, have had actual, open, notorious, exclusive and continuous possession *with or without color of title*, of any lands falling or included within the exterior boundaries of any grant confirmed or patented to any of the pueblos in this Act specified . . . shall be entitled to a decree in their favor respectively for the whole of the lands so claimed. [emphasis added]

According to Collier on September 25, 1922, "the full import of these proposals will be clear to those who know the existing condition of the lands of some of the Pueblos" (*Santa Fe New Mexican*, 1922, n.p.). At this point, Collier made it his mission to both inform the American public of this injustice and help organize the Pueblos themselves to fight the bill.

Organized by Collier and numerous Pueblo leaders in response to the threat posed by the Bursum Bill to the Pueblos—their lands, water, and by extension, entire way of life—a meeting took place on November 5, 1922, at Santo Domingo Pueblo of over one hundred representatives from all the pueblos of New Mexico. Although each pueblo governs and usually advocates for itself in most matters, a pueblo-wide council meeting to address a threat posed by a white government was not without precedent, as the Spanish would have known since the 1680 Pueblo Revolt (Bayer and Montoya, 1994; Ortiz, 1969). The *Santa Fe New Mexican* of November 4, 1922, recognized this fact, stating that "from time to time in the past all the pueblos have joined together, as in the effort to defeat a proposed law conferring citizenship on all Indians some years ago (n.p.)." However, the *New Mexican* also recognized that this meeting seemed to be of greater import than those of the past, reporting that "the present meeting, however will be the most inclusive gathering of the pueblos that has taken place for several hundred years" (n.p.). Although the *New Mexican* made only passing mention of it, the meeting was momentous in that it saw the empowerment of a pueblo-wide body that would help coordinate Pueblo efforts in land, water, educational, religious, and cultural issues for years to come. The November 6, 1922 *New Mexican* described the rebirth of the All Indian Pueblo Council (hereafter called the AIPC) this way: "The proposition for a permanent confederation of the villages has been under consideration for some time past among Indian leaders and apparently there was considerable sentiment in its favor at the meeting" (n.p.).

Of course, the presence of John Collier at this meeting has not entirely unfairly fueled the idea that it was at Collier's impetus that the council was formed. Pueblo archivist and historian Joe S. Sando (Jemez) stated, "When the Bursum Bill was introduced, John Collier, a young anthropologist who had become interested in the pueblo land question, came to a meeting to ask the pueblos to organize their council on an official basis, so that politicians in Washington would recognize them as official spokesmen for the pueblos" (Sando, 1998, 35). However, historian Kenneth R. Philp stated that it was a Pueblo man named Alcario Montoya who suggested that the Pueblos organize "as we did long ago when we drove the Spanish out" (Philp, 1981, 34). Martin Vigil (Tesuque), chairman of the AIPC from 1952 to 1957 and 1959 to 1964, confirmed that the precedent for the AIPC was set during the Spanish colonial era, adding that although he had been attending meetings since 1912, they had been going on for many years prior. At the September 2, 1929 meeting of the AIPC, Pablo Abeita (Isleta), who was governor of Isleta Pueblo in 1889 and the AIPC secretary from 1922 to 1940, stated that

> [t]he Indians have been meeting for time immemorial. I have been a member of the Council for forty years coming February, and I remember when I was a kid that the Indians would meet at the asking of any pueblo for any matters pertaining to the welfare of the Pueblo Indians. The All-Pueblo Council cannot claim having organized a Pueblo Council. (AIPC, 1929, n.p.)

Even Collier noted at the October 6, 1926 AIPC meeting that "there have been meetings other times, other years, where it was mistakenly thought, where it was said wrongly, that the meetings had been called by myself or by the organization which I represent, The American Indian Defense Association" (AIPC, 1926, n.p.). Although Collier was to exercise a profound influence on the activities of the AIPC throughout the 1920s, the council sought to maintain a distinct Indian identity in its organization, leadership, and goals.

However, the meetings after Collier began his involvement in 1922 set a new precedent for council activities and tactics. What set this new council apart from the traditional council was the idea, encouraged by Collier, that they could organize to directly influence the federal government's legislative and judicial decisions that affected the Pueblos. The initial passage of the Bursum Bill seemed to many Pueblo leaders to constitute another in a long line of attempts to deprive the Pueblos of their land. However, Collier's influence on the AIPC and his insistence that only through organization could the Pueblos overcome this skepticism is difficult to discount. Although the Pueblos themselves saw the need to organize to fight the Bursum Bill, it was Collier that helped direct many of the early activities of the AIPC. Indeed, Collier was very instrumental in orchestrating the successful campaign against the Bursum Bill. According to Vigil, it was Collier that organized both trips all over the West to raise funds and trips to Washington, D.C. to fight the bill directly. These activities of Collier and the AIPC during the 1920s and 1930s form the origins of any study of Pueblo activism in the twentieth century. It is from this beginning and from John Collier's tutelage that the Pueblo leaders learned how to organize and influence federal and state decisions that affected both their tribal sovereignty and their land and water rights in the legislative, bureaucratic, and judicial realms.

In place of the defeated Bursum Bill, Congress passed the Pueblo Lands Act of 1924 that established the Pueblo Lands Board to determine the validity of the competing (white and Indian) claims, and thus establish the true boundaries and landholdings of each pueblo. The board was made up of the secretary of the interior, the attorney general, and a member to be appointed by the president; it was given the power of subpoena, the power to take testimony, and the power to make determinations on the validity of white land claims. The proof of title had to be rather compelling: non-Indians had to demonstrate a claim under color of title predating January 6, 1902 (as shown by payment of taxes), or demonstrate a claim without color of title predating March 16, 1889 (also shown by payment of taxes). Further, if the Indian title was to be extinguished, the pueblo concerned would be reimbursed for the value of the land based on a preliminary assessment of all the pueblo grants to be made before the board's activities commenced. As a result of John Collier's appointment as commissioner of Indian affairs in 1933, the Pueblo Lands Act was amended to provide for the purchase of land and water rights for the Pueblos, as well as providing money to compensate both Indians and non-Indians who were unable to prove title to the Pueblo lands they occupied. However, according to Joe S. Sando, the 1933 act set up problems that would plague the Pueblos for years, in that it stipulated that the secretary of the interior could not make expenditures without obtaining the approval of the pueblo concerned, but at the same time, the Pueblos could only pursue acquisition of land and water rights with the approval of the secretary of the interior (Sando, 1992).

Although so much of its energy during the 1920s and 1930s had dealt with securing Pueblo lands from encroachment, the seeds for the next phase of AIPC activity, dealing specifically with the Bureau of Indian Affairs (BIA), were sown in its formative period. In the words of Pablo Abeita (Isleta) at the October 6, 1926 AIPC meeting:

It is a pity and I am very sorry that such a friction exists in Washington—that the Bureau of Indian Affairs and those gentlemen in Congress who are working for the good of the Indians should be opposed. We are here, what the white people term between the devil and the deep sea, just because the Indian office means well—our friends mean well, but they are going at each other like that—and we are between it. (AIPC, 1926, n.p.)

The BIA and congress had very different ways of dealing with Indian problems. Congress dealt with large-scale issues and multiple interest groups, but Abeita saw clearly that although Congress had much influence over vital issues of land and water, the BIA was the organization that the Pueblos dealt with on a local, daily basis. Whereas so much of the AIPC's initial activity had to do with influencing policy on a national level at the impetus of whites like Collier, local concerns were never far from the minds of the Pueblo leaders who actually made up the council. Much of this work had to do with spurring the BIA to take action either in the long term, or in reaction to emergencies that arose. The leader of the council during the first part of this period was Martin Vigil of Tesuque Pueblo. Vigil had a history of action in protection of Pueblo rights that predated the official formation of the AIPC in 1922. In 1953, Vigil was, in his own words, "drafted" to the chairmanship of the AIPC. Although the issues of the sanctity of Pueblo lands had been dealt with already, the issue of water rights in this time was becoming ever more important to the leadership of the Pueblos.

Due to the periodic droughts endemic to the region, the water situation in many of the Pueblos was desperate in 1955. Vigil led a group from a number of the northern Pueblos to a meeting in Denver with BIA officials and the commissioner of Indian affairs. He showed his attachment to local concerns, expressing frustration that the BIA was thinking in terms of ten-year programs, whereas for all he knew, there would be no water left by the time he got back to Tesuque. In the end, the BIA agreed to emergency measures to provide for water at the pueblos that needed it, drilling new wells to irrigate the fields within a week of the meeting. Although not discounting the importance of religious issues, Vigil thought that the religious leaders of the Pueblos should solve religious and cultural problems among themselves, and that the AIPC should be a forum where issues dealing with the relationships between the Pueblos and the state and federal governments and the BIA should be handled.

The first shot in the battle over protecting the scarce water supplies of the Pueblos had been fired in 1928, when the Middle Río
Grande Conservancy District (MRGCD) Act was passed. The
MRGCD was formed by the state of New Mexico in 1925 to improve
irrigation along the Río Grande. Of course, the Río Grande, being
the largest river in the state, is vital to both Indian and non-Indian irrigation, and as such is naturally one of the largest sources of contention in determining Indian water rights in New Mexico. Under the
1928 law, the Pueblos along the Río Grande (Cochiti, Isleta, San
Felipe, Sandía, Santa Ana, and Santo Domingo) have paramount
water rights (defined by those concerned as *Winters* rights) to all the
water needed to irrigate their then-irrigated lands. In addition, they
were to receive equal treatment for additional lands made irrigable by
facilities to be built by the district. In practical means, however, this
amalgam of the often-contradictory efforts of state and federal
oversight over the adjudication of the Río Grande has created an atmosphere of constant litigation over water rights that continues to
this day.

In reality, the MRGCD has had an ambiguous effect on the Pueblos.
The original contract entered into by the six Río Grande Pueblos and
the MRGCD does give the Pueblos some protection, but it does not
give the Pueblos themselves explicit jurisdiction over the water rights
guaranteed them by the original act. The MRGCD had contracted
with the Department of the Interior to protect those Indian rights
within the district. In practice, the original 8,346 acres of previously
irrigated land still have *Winters* rights, and the remainder (some fifteen thousand acres) are subject to the same regulations and adjudication methods as non-Indian lands (DuMars, O'Leary, and Utton,
1984). This may seem like a fair enough compromise, but consider-
ing that the *Winters* doctrine has been construed by some as applying
to all Indian water claims, it is far short of what the Pueblos wanted.
Of course, the groundbreaking *Arizona v. California* decision, which
clarified *Winters* and set criteria for Indian water claims, had not yet
been made.

Inclusion in the MRGCD by proxy extended the Pueblos into the
realm of water rights adjudication in the West as a whole, and not
just within New Mexico. The Río Grande, of course, flows through
Colorado, New Mexico, and Texas (besides Mexico's often over-
looked claims) and forms a part of a larger Western watershed. From

the late 1930s until the present, the scarce waters of the West have been tied up in legislation and litigation that has often either over-looked or been to the direct detriment of Indian water claims. Sando has pointed out that most of the water agreements among Western states have included language guaranteeing that the agreement shall in no way affect Indian water rights, but in practical politics, state legislatures have often passed legislation and state courts have made decisions related to these agreements that ignore Indian interests (Sando, 1992). The AIPC itself has gone so far as to state that "the agreement to include the Pueblo tribes of the Middle Rio Grande Valley in the Middle Rio Grande Conservancy District served only to limit the water rights of the tribes involved" (AIPC, 1972, 42). By including the Pueblos in the MRGCD, the federal government had contracted away their role as the protector of Indian water rights to a subsidiary agency of the state of New Mexico.

Both the Pueblos' involvement in the wider world of Western wa-ter politics and their suppression by state authorities can be seen in the case of *Texas v. New Mexico* (344 US 906, 1952). Although the Pueblos were not direct litigants in the case, it involved their inter-ests in some very important ways. In 1952, the state of Texas, upset over New Mexico's (and, by extension, the MRGCD's) plans to build the Middle Río Grande Project in the late 1940s that would increase New Mexican usage of Río Grande waters, filed a suit that was to fi-nally determine the amount of Río Grande water that each state in-volved in the Río Grande Compact was entitled to. In the Supreme Court's 1952 decision, it directed that a special master be appointed by the court to determine the amount of water to be awarded to the state of Texas, the state of New Mexico, and the MRGCD. The origi-nal opinion, however, left the question open as to whether the United States was an "indispensable party after evidence," pertaining to their role of protecting the rights of the Pueblos concerned. Be-cause the United States, in the midst of the era of termination, re-fused to involve itself in the case, another opinion was rendered in 1957. This final opinion (*Texas v. New Mexico*, 352 US 991, 1957) dismissed any further litigation in the case "because of the absence of the United States as an indispensable party" (AIPC, 1972, 49). The United States had abandoned its role as Indian advocate, in the opin-ion of the AIPC, because "there would not be enough water left for the development of large non-Indian interests." Regardless of the reasons for their actions, the U.S. government had reneged on its obligation to the Pueblo nations. The lessons learned by the Pueblos would affect the ways that both the AIPC and the individual Pueblos pursued the resulting litigation for the next thirty years.

THE AIPC AND PUEBLO WATER RIGHTS (1960–1989)

> There has been a lot said about the sacredness of our land which is our body; and the values of our culture which is our soul; but water is the blood of our tribes, and if its life-giving flow is stopped, or it is polluted, all else will die and the many thousands of years of our communal existence will come to an end.
>
> —Frenk Tenorio, San Felipe Pueblo, 1975

After their reorganization in 1922, the AIPC addressed numerous issues of concern to the Pueblos, and among these have been communication between the Pueblos, education of the Pueblo citizens, and as I address here, Pueblo land and water rights. The issue of Indian water rights is incredibly complex, and the varying legal status of the Pueblos only increases that complexity. Yet in an arid climate such as New Mexico, there is no denying the importance of water in both practical and social terms. The AIPC has explicitly stated this multi-faceted connection many times over the past seventy-six years. For example, in a 1972 report to the U.S. Commission on Civil Rights, the AIPC stated:

> The Pueblo Tribes of New Mexico have depended on the Rio Grande to sustain their lives for thousands of years. To the Pueblo Indian, the Rio Grande is a living part of the balanced scheme of nature, with which the tribes maintain a close relationship. Thus, the river is a part of the very life and existence of the Indian. When the river dies, so does the Indian. (AIPC, 1972, preface)

But to give ample background to the discussion of the AIPC's role in negotiating Pueblo Indian water rights in the twentieth century, it is essential to understand the context in which the AIPC operated during these years of change for both Indian Country and the United States as a whole.

The debate between traditional and constitutional forms of government in the Pueblos continued into these years. At the beginning of this period, only three of the pueblos had adopted the constitutional form of government advocated in the Wheeler-Howard Act. Santa Clara Pueblo was the first, approving their constitution on December 20, 1935. Isleta followed on March 27, 1947, and Laguna on November 10, 1958. The rest of the pueblos continued to use their unwritten traditional forms of government. In 1965, however, the AIPC itself adopted a constitution that spelled out its structure and the limits on its powers. Each pueblo that paid annual dues cast one vote in all matters before the council. The council took on the responsibilities of employing legal counsel, negotiating with governments, promoting or conducting the educational, health, or publicity campaigns for the member Pueblos, and promoting and fostering programs for the benefit of any or all of the member Pueblos.

What distinguishes the period from 1965 to 1989 in Indian history in general is the rise of the American Indian Movement (AIM) as well as the growing power of the sometimes opposing force of tribal governments. Activist groups like AIM, whom the AIPC usually opposed, focused on the use of the tactics of the civil rights movement such as mass demonstrations and sit-ins to gain national attention for their cause. In contrast, the AIPC mainly kept their focus on water rights and other locally important issues to the pueblos. Direct advocacy with the federal government, which had been so much a part of the council's mission after the 1920s, continued to be a major part of the AIPC's activities. Education, which had always been important to the AIPC, also took on a new emphasis during these years. Importantly, though, the AIPC began using new tactics both to educate their own people on important issues and to spread their views on issues affecting the Pueblo nations to the non-Pueblo communities. In mid-1973, the AIPC began publication of *New Mexico's 19 Pueblo News*, a monthly newspaper in which the AIPC leadership was able to raise awareness of bills and court decisions effecting the Pueblos, and of the educational, social, and alcohol rehabilitation programs run by the AIPC. The operation of a radio station also added further to the reach of the council into the widely dispersed Pueblos of New Mexico.

This type of communication proved necessary, as the late 1960s and the 1970s saw quite a few important issues come to a head for both the AIPC and the Pueblos as a whole. The adjudication of the Río Grande and other rivers continued to be a source of contention between the Pueblos and the state of New Mexico. Many in the

161

Pueblos became frustrated with the BIA and the federal government as a whole due to their reluctance to protect Pueblo water rights to the extent allowed under the *Winters* and *Arizona v. California* decisions. By the mid 1960s, the state of New Mexico was insisting that its own water laws, based on the Prior Appropriation doctrine, held sway over the Pueblos, and that the laws of New Mexico were superior to the Pueblos' *Winters* rights. This basic disagreement over the amount of water that the Pueblos could claim would become an over-thirty-year battle known as *New Mexico v. Aamodt* (537 F.2d 1102, 10th Circ., 1976).

Given the ambiguous nature of American Indian water rights in general (what Norris Hundley, Jr. in the title of his 1978 article called "confusion elevated to principle"), and the even more complex nature of Pueblo water rights, it is not surprising that the extent of Pueblo water claims would have to be decided in the judicial realm. What is surprising is that it was not until 1966 that a case was filed in this matter, that it took ten years for a decision to be made in the case, and that remnants of the original case are still being litigated to this day. Interestingly, the suit was initiated not by the Pueblos themselves, but rather by the engineer of the state of New Mexico. Also, the original suit not only dealt with the adjudication of Pueblo water rights, but included approximately one thousand non-Indian defendants (although much of the litigation has been divided between Indian and non-Indian decisions). The main issue that the state sought to argue in the case of *New Mexico v. Aamodt* (537) was that the Pueblos were subject to state water law under the doctrine of Prior Appropriation rather than the reserved rights doctrine applied to Indian tribes under the *Winters v. United States* (207) decision.

In 1969, after the original decision at the state court level had determined that the Pueblos were to be subject to New Mexico's Prior Appropriation water laws, the AIPC along with friends like Dr. Sophie B. Aberle and William Veeder began to compile legal materials for the appeal to the U.S. Court of Appeals for the Tenth Circuit. In a report written by Daniel M. Rosenfelt of the University of New Mexico School of Law, the AIPC asserted that neither the BIA nor the federal government as a whole had been sufficiently aggressive in protecting the rights of the Pueblos to the waters of the Río Grande. Also, the report argued that the loss of the Río Grande waters affected not only the Pueblos directly, but others as well, because those along the Río Grande would have to draw water from sources already being used by other Pueblos. Furthermore, the report stated that the loss of water rights amounted to attempted de facto termination of the Pueblos as "viable, permanent, and independent communities." Finally, the report condemned the state of New Mexico for its complicity in working out a settlement with the non-Indian defendants that worked to the exclusion of Pueblo interests (Rosenfelt, 1969).

The Pueblo arguments against their being subjected to New Mexico water laws were multifaceted and substantial.

- The Pueblos rebutted the state's argument that the Pueblo Lands Act limited the water rights of the Pueblos concerned. The Pueblos pointed out that water rights were explicitly not included in the language of the act.
- The Pueblos argued that the interest of the United States in Pueblo water rights did not emanate from state law and, in fact, predated it.
- The Pueblos argued that the United States had an inherent power to reserve the waters concerned and this right had never been subjected to state law.
- The Pueblos argued that Indian tribes, as wards of the government, have "immunity of the sovereign" from suit.
- The Pueblos argued that United States Supreme Court cases had given precedent to the idea that where there is any doubt as to an issue affecting Indian rights, the doubt should be resolved in favor of the Indians (Rosenfelt, 1969).

Another interesting argument was pointed out by Gregory Pease (1969). He noted that the New Mexico state constitution apparently disclaims any state jurisdiction over Indian lands. Article XXI, Section 2 reads:

The people inhabiting this state do agree and declare that they forever disclaim all right and title to . . . all lands lying within said boundaries owned or held by any Indian or Indian tribes, the right or title to which shall have been acquired through the United States, or any prior sovereignty; and that until the title of such Indian or Indian tribes shall have been extinguished the same shall be and remain subject to the disposition and under the absolute jurisdiction and control of the Congress of the United States.

However, Pease also asserted that this exclusion was in a rather tenuous position at the time. Public Law 280, which had been passed in 1953 as a part of the general move toward termination of Indian services, appeared to Pease to give the states the right to remove any constitutional impediments to asserting their jurisdiction, and also that the Indian Civil Rights Act of 1968 did nothing to alter this constitutional provision. However, Public Law 280 explicitly stated:

> Nothing in this section shall authorize the alienation, encumbrance, or taxation of any real or personal property, *including water rights*, belonging to any Indian or any Indian tribe, band, or community that is held in trust by the United States or is subject to a restriction against alienation imposed by the United States. (emphasis added)

The Pueblos in general, and the AIPC in particular, would continue to argue that Indian water issues were rightly in the domain of the federal government. However, given this understanding, the AIPC in the early 1970s pointed out in some very visible ways that the federal government's efforts to protect those water rights had been remiss.

In November 1972, with the *Aamodt* litigation still pending on appeal, the AIPC prepared a report for the U.S. Commission on Civil Rights called *The Right To Remain Indian: The Failure of the Federal Government to Protect Indian Land and Water Rights.* This report eloquently documented the relationship between the federal government and the Pueblos, including the land rights struggles of the 1920s. Probably the most important part of this relationship relating to the water struggles of the 1970s was the inherent contradiction between the role of the United States as trustee for the tribes and its responsibilities in connection with water development for non-Indians. The report then pointed out the legal responsibilities of government as trustee to provide water to the Pueblos in accordance with the guidelines set out in the *Winters* (207) and *Arizona v. California* (373) decisions. The AIPC asserted what was true in legal theory, but not in practice: that the *Winters* doctrine combined with other cases was the "definitive rule upon which protection of the water rights of the tribes is based" (AIPC, 1972, 17). The AIPC then gave a brief history of the various water projects of the past fifty years and the effects of these projects on the Pueblos. Central to their argument of governmental failure were the various reclamation activities that the federal government had undertaken in the interest of white settlers and the failure of the MRGCD to provide water for Indian lands.

The AIPC made its case in the report by asserting that, given the importance of the waters of the Río Grande to the viability of the

Pueblos, that the United States had been derelict in its responsibilities by not enforcing irrigation of newly reclaimed lands by MRGCD waters, and further, that the federal government had been remiss in protecting Pueblo waters in the face of growing non-Indian demand due to the post-World War II population boom experienced in the West. The proof of this negligence on the part of the government was its refusal to involve itself as an advocate for the Pueblos in the case of *Texas v. New Mexico.*

THE LEGACY OF THE AIPC AND THE FUTURE OF INDIAN WATER

> The matter of Pueblo Indian water rights has been in and out of the courts for so long that one more case would appear to indicate nothing less than extreme harassment on the part of the state of New Mexico against the Pueblos.
> —Joe S. Sando, Jemez Pueblo, 1992

Water continues to be a vital issue with the Pueblos. Ancillary cases of the *Aamodt* litigation involving Acoma, Zia, Santa Ana, and Jemez Pueblos continue to proceed back and forth through the federal court system. According to a 1999 interview with James Hena, former governor of Tesuque Pueblo and former AIPC chairman, some pueblos like Tesuque are not able to maximize their land under irrigation for lack of water. The growth of cities like Santa Fe and its suburbs have led to the tapping of the groundwater underlying the Tesuque land. Outside issues such as casino gambling threaten to divide the AIPC between the haves and have-nots. According to a 1999 interview with Roy Montoya, secretary of Santa Ana Pueblo, the gambling issue has ramifications for Pueblo water issues because those who have gone the gambling route have additional resources that they are able to devote to water development. Rivalries between pueblos over where water facilities will be built also have divided many pueblos.

However, there are hopeful signs for the Pueblos. Tesuque Pueblo, acting on its own, was able to stop the Santa Fe Ski Bowl from expanding into Tesuque Canyon and further infringing on Tesuque's ecosystem. Overtures between some Pueblos and the Bureau of Land Management (BLM) regarding joint management of lands raise the possibility of the idea of entire ecosystem management and cooperation finally coming to fruition. The AIPC continues to exist, although today it concentrates more on issues of education and alcohol-addiction programs. It has served a vital role in protecting and maintaining the Pueblos' land base and water rights over the twentieth century. Its example and that of other pan-tribal groups have shown that successful coordinated action can be taken to protect tribal self-determination over land and water resources.

REFERENCES

All Indian Pueblo Council, Inc. October 6, 1926. *Minutes*. S. Lyman Tyler Collection, Western Americana Library, University of Utah Special Collections, Salt Lake City (hereafter called SLTC).

———. September 2, 1929. *Minutes*. SLTC.

———. 1972. *The Right To Remain Indian: The Failure of the Federal Government to Protect Indian Land and Water Rights*. Albuquerque, NM: All Indian Pueblo Council, Inc.

Bayer, Laura, with Floyd Montoya (Santa Ana) and the Pueblo of Santa Ana. 1994. *Santa Ana: The People, the Pueblo, and the History of Tamaya*. Albuquerque: University of New Mexico Press.

Deloria, Jr., Vine, and Clifford M. Lytle. 1983. *American Indians, American Justice*. Austin: University of Texas Press.

———. 1984. *The Nations Within: The Past and Future of American Indian Sovereignty*. Austin: University of Texas Press.

DuMars, Charles T., Marilyn O'Leary, and Albert E. Utton. 1984. *Pueblo Indian Water Rights*. Tucson: University of Arizona Press.

Fixico, Donald L. 1998. *The Invasion of Indian Country in the Twentieth Century: American Capitalism and Tribal Natural Resources*. Niwot: University Press of Colorado.

Hundley, Jr., Norris. 1978. The Dark and Bloody Ground of Indian Water Rights: Confusion Elevated to Principle. *Western Historical Quarterly*, 9: 454-82.

———. 1982. The "Winters" Decision and Indian Water Rights: A Mystery Reexamined. *Western Historical Quarterly*, 13: 17-42.

Lewis, David Rich. 1995. Native Americans and the Environment: A Survey of Twentieth-Century Issues. *American Indian Quarterly*, 19, no. 3: 423-450.

Lowie, Robert H. 1954. *Indians of the Plains*. Lincoln: University of Nebraska Press.

McCool, Daniel. 1987. Precedents for the Winters Doctrine: Seven Legal Principles. *Journal of the Southwest*, 29 no. 2: 164-178.

———. 1994. *Command of the Waters: Iron Triangles, Federal Water Development, and Indian Water*. Tucson: University of Arizona Press.

Newville, Ed. 1989. Pueblo Indian Water Rights: Overview and Update on the Aamodt Litigation. *Natural Resources Journal*, 29: 251-278.

Ortiz, Alfonso (San Juan). 1969. *The Tewa World: Space, Time, Being, and Becoming in a Pueblo Society.* Chicago: University of Chicago Press.

Pease, Gregory. 1969. Constitutional Revision-Indians in the New Mexico Constitution. *Natural Resources Journal,* 9: n.p.

Philp, Kenneth R. 1981. *John Collier's Crusade for Indian Reform, 1920-1954.* Tucson: University of Arizona Press.

Rosenfelt, Daniel M. 1969. *Report on the Protection of Pueblo Indian Rights to the Use of Water in the Rio Grande Basin: A Discussion of Pending Litigation.* Albuquerque: University of New Mexico School of Law.

Sando, Joe S. 1992. *Pueblo Nations: Eight Centuries of Pueblo Indian History.* Santa Fe, NM: Clear Light Publishers.

———. 1998. *Pueblo Profiles: Cultural Identity through Centuries of Change.* Santa Fe: Clear Light Publishers.

Santa Fe New Mexican, September 20, 1922. "The Bursum Indian Bill: Full Text of Measure Vitally Affecting Fate of New Mexico Pueblos, Now Pending in Congress."

———. September 25, 1922. "Effect of Indian Bill Would Be to Cripple, Destroy Some of Pueblos, Says Collier."

———. November 4, 1922. "All Pueblos To Send Men To Conference Held Sunday."

———. November 6, 1922. "Indian Bill Will Destroy Pueblo Life, Say Indians In Memorial To Country."

Weber, David J. 1982. *The Mexican Frontier 1821-1846: The American Southwest Under Mexico.* Albuquerque: University of New Mexico Press, Albuquerque.

———. 1992. *The Spanish Frontier in North America.* New Haven, CT: Yale University Press.

Wilkinson, Charles F. 1987. *American Indians, Time, and the Law: Native Societies in a Modern Constitutional Democracy.* New Haven, CT: Yale University Press.

Wilson, Angela Cavender. 1998. American Indian History or Non-Indian Perceptions of American Indian History? In *Natives and Academics: Researching and Writing about American Indians.* Devon A. Mihesuah, (ed.). Lincoln: University of Nebraska Press.

Worster, Donald. 1985. *Rivers of Empire: Water, Aridity, and the Growth of the American West.* New York: Oxford University Press.

Ten Haunted by Waters: Water Rhetorics as Conservation Politics in Films of the American West

Joan M. Blauwkamp

The struggle over water has been among the most contentious issues in the politics of the American West. Los Angeles, Phoenix, and Las Vegas struggle to meet the increasing demands brought by population explosions in sites with desert ecologies. Conservationists try to use water limits to constrain growth in urban areas and to protect the ecosystems and smaller communities threatened by overdrawing of aquifers, transportation of water resources from other areas, and development. Throughout the West, nations, states, and local communities compete with one another for shares of common water resources and rights to scarce water supplies. In the Great Plains, surface pollution from agricultural runoff, stream flows in the Platte and other rivers, and aquifer pumping are the subjects of heated debate, legislation, and most recently, litigation.

It should be no surprise that water has figured prominently in western storytelling, including films. Sociologist Kiku Adatto writes, "Popular movies derive their appeal by engaging us in familiar narratives. For all the different stories they tell, they often reiterate certain myths and ideals that resonate with our collective self-understanding" (1993, 125). The collective self-understanding of the American West is defined by its aridity and its people by their search for reliable sources of water (Reisner, 1993; Stegner, 1992; Thomas, 1991). The search for "reliable" sources of water suggests that the concern of western residents should be not only with supply but also with sustainability and, therefore, with conservation of water. This chapter addresses the strategies for conservation politics suggested by three recent American films about water: *The Milagro Beanfield War*, *Thunderheart*, and *A River Runs Through It*.

Roman Polanski's classic 1974 film *Chinatown* set the standard for film treatments of water politics in the West. Although not the first film to address water issues (see the 1953 western *Shane,* for example), *Chinatown* is arguably the most prominent film about water politics set in the contemporary political scene. *Chinatown,* with screenplay by Robert Towne, told a liberal-democratic story of water politics. As the film ends, Los Angeles will build itself a new dam, even though the site is unstable. Wealthy businessman Noah Cross and his cronies will succeed in buying up land northwest of Los Angeles, then incorporating the land into the city limits so they can irrigate it with subsidized water from the new dam. Noah Cross succeeds with his scheme to build a so-called public works project for his own private enrichment because he is already rich. And as protagonist Jake Gittes summed it up: "The rich can get away with anything." *Chinatown* portrays water politics as a competition among interests, and in this competition, the deck is heavily stacked in favor of the wealthy. Money is power, and Noah Cross has enough to get what he wants.

This is not an encouraging tale for conservation-minded citizens. Social movements such as the conservation movement in the United States generally lack the financial resources to rival corporate developers or wealthy industrialists. Westerners have a saying, "water flows uphill to money" (the expression is quoted in Brown and Ingram, 1987 and in Reisner, 1993). If this adage is true, the possibilities for conservation seem bleak. Fortunately, money does not appear to be the sole key to political influence. Several recent films about water politics in the West explore the possibilities for conservation on republican terms, with much more hopeful results.

SITUATING CIVIC REPUBLICANISM

In the liberal-democratic view, the political realm is a contest in which political actors are competitors jockeying for the power to successfully promote their own self-interests; Hariman (1995) calls this approach to politics the "realist" style. By contrast, civic republicanism takes the view that the political realm rewards skilled orators who are able to construct consensus around political goals through their rhetorical abilities and their embodiment of civic virtue. In this

169

view of politics, victory belongs not to the biggest fortune but to the greatest skill, skill at constructing the symbolic and rhetorical situation to favor one's cause.

Robert Hariman's excellent analysis of republican political style in Cicero's letters identifies three key features of a republican approach to politics: 1) public discourse as the key mode of political influence, 2) consensus as the means and end of governance, and 3) personal embodiment of civic virtue as the model of leadership (1995). Implicit in Hariman's analysis and more clearly identified elsewhere is a fourth feature of republican politics: an emphasis on the public space as the locus of political action (Arendt, 1958; Kemmis, 1990).

It is this fourth feature of republican politics that creates such an affinity between civic republicanism and conservationism. Both ideologies recognize the importance of common ground to political action. For civic republicans and conservation-minded citizens, that common ground creates what Daniel Kemmis has called a "politics of inhabitation," a recognition that we share our place with others, and so we must come to a shared understanding of what should be done there in order to live well together (1990; see also Snyder, 1995). A politics of inhabitation implies that within a community there are agreed-upon ways of doing things, "habituated patterns of work, play, grieving, and celebration designed to enable people to live well in a place" (Kemmis, 1990, 80). Those "habituated patterns" that Kemmis describes are better known to civic republicans as "practices," activities that tie an individual to a community, to its traditions, and to its shared values (MacIntyre, 1984). Conservationists like Gary Snyder (1990), Wendell Berry (1977), and Michael Pollan (1991) use the term to argue that practices connect a person to his or her place, and that only the people who are part of a place understand its potentialities and its limitations. They are, therefore, the only people who are able to care for it well.

These connections between community and place and between place and practices are enacted in the three films that are the subjects of this chapter. Each story exemplifies a republican approach to water politics, and in so doing, each creates an environment in which conservation is possible.

The Milagro Beanfield War (1988)

In the town of Milagro, New Mexico, a miracle occurs. The town rediscovers its sense of community by restoring its connection to its water. This is miraculous, for when the credits roll on the opening sequence of Robert Redford's film, the battle in liberal-democratic terms has already been lost. Wealthy developer Ladd Devine has already obtained the rights to the water flowing through Milagro Valley. He has the necessary zoning permits, the support of the state government, everything that he needs to begin construction on the new recreation area he has planned. All the legal and institutional hurdles have been cleared. But this is the beginning, not the end of the story.

Milagro's local mechanic Ruby Archuleta tries to fight Devine using standard liberal-democratic political moves. She tries to form the local residents into a public interest group, the Milagro Land and Water Protection Association, but she cannot get the people of Milagro to agree on what is to be done. She tries to approach them individually by circulating a petition, but no one wants to sign. They know already that poor people cannot thwart the plans of rich people by signing papers. Retired activist lawyer and current newspaper publisher Charlie Bloom exemplifies their resignation when he tells Ruby: "These people have been burned too many times signing papers. You will just get their hopes up, then they'll come crashing down and blame it all on you." Liberal-democratic politics advantage the wealthy; the people of Milagro cannot compete in that arena.

Enter Joe Mondragon. Mondragon is desperate for work. He tries to get a construction job at the new development but is turned away. His options are to farm his family's land in Milagro or to become a migrant worker and leave Milagro. His refusal to leave his home leads Joe to a desperate act. He decides to farm his family's field, growing pinto beans, and to take the water he needs to irrigate, despite the restrictive water laws that have turned Milagro's once fertile fields to dust.

This act of defiance is initially a personal one for Mondragon; he complains at Ruby's town meeting that he did not start his bean field so that the town "could form a committee." On liberal-democratic terms, his struggle is also futile. As Bloom tells Ruby just after Joe begins plowing: "I've been representing guys like Joe Mondragon for most of my life. I know a lost cause when I see one." Despite this dire pronouncement, Joe's actions become a catalyst that eventually brings the community to a consensus in opposition to Devine's recreation area.

Joe Mondragon embodies civic virtue. He shows courage in the face of intimidation and violence from Devine's men. He becomes the symbol of Milagro, and the townspeople take action to protect him and his bean field. The old men in town defend him from harassment by the local Forest Service officers in Devine's pocket. Bloom, the newspaperman, opens the sluice gate to irrigate the bean field when Joe is on the run from Kyril Montana, who was sent from the capital to "discourage" Joe from growing the bean field without making an arrest that would draw attention to the Governor's questionable dealings with Devine. The "posse" assembled by Montana protects Joe from capture by making sure they do not find him. As Montana complains to Milagro's Sheriff Montoya: "This posse couldn't find itself." Sheriff Montoya responds, "Unless it wanted to." Even Devine's foreman intervenes to protect Joe from being shot by Montana.

Joe's example spurs other acts that undermine Devine's apparent advantage. An anonymous saboteur begins a series of nightly pranks on Devine and the recreation area. All these acts appear, in the end, to be the work of Don Amarante Cordova, the oldest man in Milagro, alone or in conjunction with the other elderly men who defended Joe from the Forest Service officers. However, the confessed perpetrator is "El Brazo Onofre," referencing a local legend about activities attributed to the severed arm of a Milagro elder, Onofre Martinez. Onofre's arm is believed to "lead a life of its own." By linking the sabotage to "El Brazo," the assaults against Devine and the recreation area appear to come from Milagro itself. They cannot be dismissed as the acts of a radical individual or minority group. The very arm of the community appears to have risen up in its defense. This powerful symbol helps to undermine community support for the recreation area.

The fight against Devine takes another step forward when a "miracle" occurs that suggests Nature itself opposes Devine. On a suspicion that newspaperman Charlie Bloom will use Joe's bean field to generate negative publicity about the recreation area, Devine sends his henchman to follow Bloom on his deliveries, buy all copies of his paper *La Voz del Norte* (the Voice of the North) and burn them. A sudden burst of wind takes *La Voz* from its pyre and deposits the copies on the streets of Milagro. The Voice of the North falls from the sky, lending divine support to the paper's warnings about the recreation area.

The battle of symbols further turns against Devine when the prankster/"El Brazo" begins to write notes that refer to Devine as the *zopilote,* the buzzard or vulture. Once this appellation catches on, the recreation area can no longer be characterized as "something good for this community," as Devine would have liked. Instead, he is a predator, waiting for the town of Milagro to die so he can feast on its carrion. The label brings home Ruby and Bloom's warnings. If the recreation area comes, it means the death of Milagro. Devine is not the town's benefactor, but its scavenger.

By contrast, Joe's virtue is on display in his bean field. As MacIntyre observes, farming is a practice, and as such, it "involves standards of excellence and obedience to rules as well as the achievement of goods" (1984, 190). The bean field provides food for Joe's family and crops to sell, but it also demonstrates Joe's character, and thereby his worthiness to lead. Milagro's old-timers come to Joe's defense so readily because they have observed his bean field and judged it admirable. Brown and Ingram (1987) note the importance of farming to the culture and traditions of northern New Mexico Hispanics, even if it is only a part-time activity, as it is for Joe Mondragon in this story. Farming is not the sole means of economic support for the Mondragon family, but it is central to the family's traditions. We are made aware of this as word spreads quickly through the town that Joe is "irrigating *his father's* bean field" (my emphasis).

Joe's return to farming connects him to his land but also to his past. The people of Milagro have deep roots in their place. Don Amarante tells Herbie Platte, the sociology student from NYU who is studying "indigenous cultures of the American Southwest," that his people have been growing beans in that field since the Treaty of Guadalupe Hidalgo. The reference to the 1848 treaty can be viewed as a complaint against the change in water rights that precipitated this crisis. The Treaty of Guadalupe Hidalgo included a protocol in which the U.S. government agreed to recognize water rights granted under Spanish or Mexican rule (Brown and Ingram, 1987). Don

Amarante seems to be suggesting that the citizens of Milagro have a prior appropriation right to the water, and that by siding with Devine, the government is breaking the agreement it made in the treaty. Likewise, Ruby tells Bloom that her people "have a right to stay" since they have been in Milagro for three hundred years. Amarante and Ruby's commitment to stopping the development lies in their awareness of the vital connection between Milagro and its people, and the role of water in preserving that connection.

Joe's bean field represents the revival of the community of Milagro. It cannot survive, let alone flourish, without water. Neither can Milagro. As the bean field changes from lifeless dust to healthy green, the community itself acquires a renewed sense of unity and purpose. The entire town gathers to help in the harvest of the bean field. The event becomes a symbolic founding of the town, a dramatic moment that reconstitutes the commitment of the people to each other and to preserving their place in the world (Hariman, 1995). Their renewed unity and dedication are strong enough to convince New Mexico's governor to withdraw his support for the recreation area.

Despite his financial advantage, Devine is forced to abandon his plan for a recreation area. Milagro pulls off its miracle, not on legal grounds, but because it is able to reconfigure the symbolic and rhetorical situation to favor its cause. Joe's leadership constructs a renewed consensus and commitment among the people of Milagro to live well together in their place in the world. To do so, they must stop the recreation area, and they do.

The Milagro Beanfield War turns the political tale of *Chinatown* on its head. Both stories begin with a wealthy and powerful man intent on achieving his own interests at the expense of the community. In *Chinatown*, the wealth and power of Noah Cross was insurmountable. But in Milagro, the rich could not "get away with anything." The liberal-democratic approach of forming an interest group failed to stop Devine, but by mobilizing the citizens of Milagro through their symbolic connections to their land and the water that nourishes it, the recreation area was blocked.

This story affirms the importance of water to western communities. The people of Milagro did not understand the water laws that prevented them from irrigating their land, but they understood clearly that water was the lifeblood of their valley—it should be used to irrigate their crops, not Devine's golf courses. The symbol of Joe's green, flourishing bean field became the catalyst that brought the community together against Devine. Hence the film suggests an alternative approach to interest-group lobbying and litigation for the conservation-minded citizen concerned with water politics. Many communities may not have a trickster spirit who controls the winds, but their traditions, customs, histories, and connections to their place in the world have symbolic resonance for residents. For northern New Mexico Hispanics, the desire to preserve their culture is vitally connected to water. Without water, the land cannot be farmed. Without farming, the community cannot support itself, families break apart, and the culture, language, and traditions of the people are lost (Brown and Ingram, 1987). Other western communities, including the Tohono O'odham Nation in southern Arizona, attach similar symbolic importance to water (Brown and Ingram, 1987). By drawing on those ties that bind a community together, it may be possible to develop a consensus within the community that can challenge the interests of the most wealthy corporation or industrialist. Symbolic politics can be used to develop community consensus regarding the need for water conservation and ecologically responsible use.

THUNDERHEART (1992)

On the Bear Creek Indian Reservation in the Badlands, South Dakota, the water is contaminated. Children are getting sick. Ranchers are finding stillborn calves among their stock. This would seem to be a cause for government action, but instead the United States sends in the FBI to investigate one murder, a murder in the middle of a war zone.

The setting for Michael Apted's film is loosely based on events that took place on Indian reservations in the 1970s, particularly the sometimes violent conflicts between progovernment Indians and the supporters and members of the American Indian Movement, here called the Aboriginal Rights Movement (ARM for short). The progovernment faction controls the tribal council on the Bear Creek Reservation. Tribal President Jack Milton dominates the reservation with an iron fist, using a band of armed thugs called the Guardians of the Ogallala Nation (or GOONs) to harass, assault, and even kill ARM

175

supporters. ARM member Maggie Eagle Bear documents sixty-one ARM supporters whose deaths were never investigated. Given these circumstances, it is significant to note that the FBI is called in to investigate when a progovernment Indian is murdered.

The FBI agent sent to investigate the murder is Raymond Levoi, whose father was half-Indian, a Sioux, but who died when Ray was seven years old. Out of shame over his father's alcoholism, Ray has systematically denied his identity as an Indian. Nonetheless, he is sent to Bear Creek Reservation as "an Indian FBI agent," because the FBI hopes the gesture of sending the Indians "one of their own" will ease tensions and increase cooperation during the murder investigation. Once there, Ray begins a journey to find himself and the truth about the murder he was sent to investigate.

Levoi's superior is a legendary FBI man named Frank Coutelle, whose investigation of the murder targets one of the last surviving ARM members, Jimmy Looks Twice. The murder victim, progovernment tribal council member Leo Fast Elk, is found lying out on the sandstone hills of the Badlands with the ARM symbol, an eagle feather inside a circle, marking the scene. Tribal cop Walter Crow Horse is conducting an independent investigation into Leo's murder, despite the FBI's jurisdiction in the case, and he eventually opens Ray's eyes to Jimmy's innocence. Crow Horse discovers that Leo was killed on the property of ARM member Maggie Eagle Bear, where it abuts the Little Walking River. He was transported to the Badlands site in his own car, which is later found (by Crow Horse) partially submerged in the river. Leo's body had been moved to divert attention from the connection between the river and his murder.

With the help of Crow Horse and tribal elder Samuel Reaches, Ray discovers the true motive for Leo Fast Elk's murder. Despite opposition from the tribal council, FBI agent Frank Coutelle and Tribal President Jack Milton had brokered a land deal to drill for uranium on reservation land, a spot called Red Deer Table, which had been the traditional home of the tribe before the government forced them to move. Because Red Deer Table is the headwaters of the Little Walking River, the test drilling caused the river's contamination. To prevent exposure, Coutelle and Milton had Leo killed and framed Jimmy Looks Twice to destroy ARM, the only people who might be able to stop the mining operation from going forward.

The contamination of the Little Walking River threatens the lives of everyone on the reservation. It also threatens the life of the tribe. The tribe exists in its connection to the land. This connection is repeatedly affirmed in the film. Jimmy describes ARM's struggle to FBI agents Coutelle and Levoi as "a five-hundred-year-old resistance." Maggie credits Jimmy's escape from FBI custody to his ability to "shape shift into different animals." Indeed, a buck appears as if from nowhere at the site where Jimmy was last spotted by Agent Levoi. Walter Crow Horse tells Ray that Grandpa Reaches learned of Leo Fast Elk's death from an owl. Later, when Ray's connection to his tribal heritage is reestablished, an owl appears to tell him of another death. When Ray asks Jimmy why he is considered such a threat, he replies: "There's a way to live with Earth and a way not to live with Earth. We choose the way of Earth."

Red Deer Table is more than just the headwaters of the river; it is "the source." Literally, it is the source of the reservation's water supply, but mythically, it is the source of the people themselves. Both Maggie Eagle Bear and Walter Crow Horse use this term to refer to Red Deer Table. Grandpa Reaches makes the same point, but he is a bit more cryptic; he tells Ray to "listen to the water" to find the truth about Leo's murder and about Ray himself.

In the course of his investigation, Ray discovers his true identity. In a vision, he sees himself chased down by a cavalry officer, running through a field with other Indians, mostly women and children. He is shot in the back. Later he recognizes the place of his vision, the Wounded Knee Memorial. Grandpa Reaches interprets the vision for him: He is descended from a holy man who died at Wounded Knee, named Thunderheart, with whom he shares the same spirit. Ray is literally the embodiment of this ancient leader, which confirms Ray's connection to the Indians of the Bear Creek Reservation and establishes his qualifications to lead. With the testimonial of Grandpa Reaches, Ray becomes more than just a derided "Washington Redskin." Instead, "Thunderheart has come. Sent here to a troubled place to help his people."

Like Joe in *The Milagro Beanfield War,* Ray helps his people by constructing a new consensus within the community. He can only succeed when he acknowledges that legal procedures are insufficient to save Jimmy or expose Milton and Coutelle. He tries to gather evidence that exposes Richard Yellow Hawk, a parolee planted in ARM as a spy by Coutelle, as the real murderer; but his evidence cannot compete against Coutelle's fabricated evidence against Jimmy, which he has built into "an airtight case." His frustration mounts as Maggie dismisses his evidence: "That's not power, Ray, that's paper. Power is a rainstorm. Power is that river, right there. And that is what I have to protect." All the paper in the world will not save the Little Walking River or Jimmy, because the Indians cannot compete in the world where paper rules (either as legal documents or money). From the point of view of the U.S. government, they are, as Coutelle notes, "a conquered people . . . whose future is dictated by the nation that conquered them." Their only means of stopping the uranium mining is to change the symbolic and rhetorical situation to their advantage.

Milton and his GOONs have been able to do as they wish with the land because the tribe has been divided, the people fighting among themselves. Ray's transformation into a leader, one willing to sacrifice himself for his people, helps to forge a new consensus. When Ray and Crow Horse are on the run from Coutelle and the GOON squad, they head for the "Stronghold," the cliff wall where Thunderheart and the other "old ones" were headed when the cavalry massacred them. By reenacting the flight of the old ones, Ray exemplifies the connection between the tribe and the land. The land is the life source of the tribe, and this is illustrated with the sound effect of a heartbeat that echoes through the scene at the Stronghold. Ray's stand at the Stronghold becomes a symbolic founding for the tribe, a reaffirmation of their unity, when Reaches and the other tribal members gathered in arms at the top of the cliff to save Ray and Walter from the GOONs. Their appearance signals a new ending to the Wounded Knee story, one in which the tribe is not massacred, but instead emerges victorious. As Ray proclaims to Coutelle and the GOONs, "This land is not for sale."

In order for the tribe to survive, the people needed to stop the uranium mining operation. They could only do so by reestablishing their connections to each other and to their place on Earth. By revisiting their history with Red Deer Table, with the site of the Wounded Knee massacre, and finally, with the Stronghold, the tribe found a new consensus and a renewed commitment to fight for their future. Ray became the symbol of the tribe; in his transformation to an Indian, he restored the people's sense of what it means to be a tribe.

Like *The Milagro Beanfield War, Thunderheart* tells the story of a poor community threatened by the development plans of wealthy and powerful men intent on pursuing their own self-interest at any cost. Again we see the significance of water for the life and health of a community. The contamination of that source of life and health is a threat to the entire community, but the community divided against itself cannot act. Only when the water is symbolically connected to the tribe's history, traditions, and identity does a consensus emerge. The community united has the strength to protect itself and its source—its water.

A RIVER RUNS THROUGH IT (1992)

Released the same year as *Thunderheart,* Robert Redford's film of Norman Maclean's autobiographical novel has a different connection to conservation politics. Maclean's story of his youth is set in the 1920s, a time when settlements in Montana were rather small and sparse. Conservation had yet to become a key concern or a political issue. *A River Runs Through It* is a much less overtly political film than either *The Milagro Beanfield War* or *Thunderheart.* Nonetheless, it is an appropriate subject for this chapter, because many of the elements of a republican style that are present in the other two films also appear in this one. Given that connection, this film also has some lessons to teach regarding political strategies for water conservation in the West.

As the people of Milagro are vitally connected to their valley, and the tribe of the Bear Creek Reservation cannot be separated from the land that is their home, the Macleans are rooted deep in Montana. Its centrality to their family identity is evident throughout the film. Early in the film, an elderly Norman Maclean (voiced by Robert Redford) remarks that in his family, "there was no clear line between religion and fly fishing." His next sentence explains this surprising statement: "We lived at the junction of great trout rivers in Missoula, Montana." Proper worship extends beyond rites within church walls; it extends as well to developing an appreciation and understanding for God's creation as revealed in the particular place where one lives. For the Macleans, that appreciation and understanding emerged from "picking up God's rhythms," in learning to fly-cast.

The ties of this close-knit family extend beyond their human members to their place; they regard the Big Blackfoot River as "our family river." Norman recollects his childhood shaped by his father's belief that education comes as much from exploring the Bitterroot Mountains as learning the three Rs. Norman courts his future wife by metaphorically linking dancing in her arms to the experience of watching the sun rise over the mountains. Norman's brother Paul determines to "never leave Montana" despite the dangerous lifestyle that eventually results in his murder. Finally, in the closing sequence, an elderly Norman muses that his memories of the people in his life that he has loved and lost are present to him when he returns to fish the Big Blackfoot River and listens to the water rushing through the canyon:

> In the Arctic half-light of the canyon, all existence fades to a being with my soul and memories and the sounds of the Big Blackfoot River and a four-count rhythm and the hope that a fish will rise. Eventually, all things merge into one, and a river runs through it. The river was cut by the world's great flood and runs over rocks from the basement of time. On some of the rocks are timeless raindrops. Under the rocks are the words, and some of the words are theirs. I am haunted by waters. (Maclean [1976], 1989)

As presented in Maclean's story, fly-fishing is a practice, in the republican sense, an act that connects the practitioner to place and tradition, an enactment of civic virtue. Norman Maclean and his brother Paul apprentice in fly-fishing under their father, the Reverend Maclean, from whom they "received as many hours of instruction in fly fishing as in all other spiritual matters." He teaches them to cast using their mother's metronome, so that they may learn to "pick up God's rhythms." For the Reverend Maclean, "all good things—trout as well as eternal salvation—come by grace. And grace comes by art, and art does not come easy." In his novel, Norman Maclean writes that he and his brother would have preferred to learn how to fish by catching fish, instead of the hours they spent learning to properly cast (Maclean, 1989). Instead, their father introduced them to fishing as a practice, in which there was discipline to be maintained, rules to be followed, and in which "nobody who did not know how to fish would be allowed to disgrace a fish by catching him." From this perspective, catching fish is not sufficient to make Norman or Paul fishermen; learning how to fish is. This is analogous to Alasdair MacIntyre's (1984) distinction between throwing a football and playing the game of football, between bricklaying and architecture, or between planting turnips and farming. In fishing as in all practices, virtue is hard work.

Norman and Paul both learn to fish, although Paul excels to become the virtuoso fisherman, who can credibly describe himself as needing "only three more years before I can think like a fish." He eventually surpasses his father's skill and finds a rhythm of his own, which he calls "shadow casting." His prowess earns Paul his brother's and father's admiration. The natural restraint of the Reverend Maclean limits his praise of Paul to a simple declarative sentence: "You are a fine fisherman." Norman goes further, describing his brother as "an artist." Paul's art earns him fame throughout that corner of the world. Even Jessie Burns, Norman's future wife, has heard of Paul Maclean and knows him as "the fishing newspaperman," although she is from the somewhat distant town of Wolf Creek. When Norman expresses surprise at his brother's reputation as a great fisherman, she remarks: "I thought everybody knew."

There was no community to mobilize or unite in conservation of the Big Blackfoot or the other trout rivers in Montana while Paul Maclean was alive and practicing his art. It was, as Norman said, "a world with dew still on it." However, that world is in need of conservation now, and it is that need that makes *A River Runs Through It* an appropriate film to include in this chapter.

The stories depicted in the other two films are stories of political struggle, communities that found renewed consensus and will through virtuous leadership and a renewed sense of connection to their place in the world. The fictional characters in those films were successful at altering their political and rhetorical situations in order to protect their water, and thereby their communities. They provide models of what might be done by conservation-minded citizens in similar real-world situations. They suggest that all liberal democracies need a dose of republican political style in order to work properly. Money is not the sole key to political influence, but it is a powerful tool that is perhaps best countered by skilled civic republicans, who understand the values of community, consensus, and virtuous leadership.

In the case of *A River Runs Through It*, Redford's initiative in making a powerful and moving film of Maclean's story has created a more favorable environment for conservation. A film such as Redford's provides a potential means for altering our perceptions of water and conservation politics in Montana and elsewhere. The story of Norman Maclean and his fly-fishing brother has entered the American mythos, particularly in Montana and the rest of the West. One sees copies of Maclean's novel for sale next to nature guides in bookstores and gift shops at tourist sites, such as Yellowstone National Park. My father has a T-shirt he bought in Missoula with the slogan that reads: "Montana: A Tourist Runs Through It." Given the popularity and acclaim of Redford's film, and its subsequent renewal of interest in Maclean's writing, the political and rhetorical situation of Montana's great fly-fishing rivers has been altered. We may be more likely to favor a Wild and Scenic Rivers designation or other protections for the Big Blackfoot, the Clark Fork, and the other rivers in Montana that have seen their value as fly-fishing sites diminished by dams and pollution. As tourism increases in importance for Montana's economy, the film provides new incentives to maintain or restore the natural state of those great trout rivers. The Bitterroot valley may no longer be a world with dew on it, but the impact of the film may be to instill in us a desire not to allow the world we saw through Norman Maclean's eyes vanish from the face of the Earth.

CONSERVATION AND CIVIC REPUBLICANISM

Most of the surface water in the West is already appropriated (MacDonnell, 1998). Groundwater is an exhaustible resource that is being rapidly depleted (Reisner, 1993). The importance of water to politics in the West is likely to grow even stronger in the future as management of dwindling supplies fails to meet demands. How will water policy be made? Will water continue to flow uphill toward money or will public interest drive water allocation and use?

The message of *Chinatown* is that water will continue to flow uphill toward money. The other films considered in this chapter provide strategies that might be used in real-world situations to enact water policies aimed at conservation in the public interest. Local participation and control of water decisions are needed for the development of such conservation policies. Local communities are best positioned as conservators of their home ecologies, including water, because they understand the possibilities, requirements, and limitations of their land and because they must live firsthand with the consequences of decisions, both good and bad (Berry, 1977; Pollan, 1991; Snyder, 1990). When a community is divided, conservationists tend to be dismissed as mere special interests or vocal minorities, usually lacking the wealth, status, or power to make their voices more than loud annoyances in the policy process. For successful conservation, community consensus is required. This point is illustrated nicely in *The Milagro Beanfield War* and *Thunderheart.* Both communities begin deeply divided, and this conflict permits the opportunism that nearly destroys them.

Consensus is difficult to develop. The public is hardly a monolithic entity; it comprises an almost unlimited variety of groups and individuals with different (and often opposing) views and values. Even an apparently homogeneous community can find it difficult to agree on anything. This situation complicates the attempt to involve the public in policymaking. Too often, policymakers pay scant attention to citizen comments at public hearings, in part because they can be dismissed as unrepresentative of the community as a whole. This dismissive attitude can result in a public that feels alienated from government and from meaningful participation in the policy process (McWilliams, 1995). Lack of community consensus complicates decisionmaking by elected representatives as well, for one is certain to displease some segment of the constituency regardless of the choice one makes. In making water policy, legislators often write rather vague policy guidelines and leave the meat of the policy to administrative units, like the Department of Water Resources and Natural Resource Districts in Nebraska. This technique allows the legislators to transform political problems into technical ones that insulate them from political fallout from unpopular decisions (Longo and Miewald, 1989). If consensus were to develop, legislators could make decisions with greater confidence of public approval, and the public would feel greater confidence in the responsiveness of government.

Rhetorical and symbolic politics have the potential to develop consensus. In *The Milagro Beanfield War,* the symbol of Joe Mondragon's bean field, the rhetorical valorization of Joe, and the vilification of Devine turned the community of Milagro against the development plans and united a previously fractured and fractious town into a potent political force. In *Thunderheart,* a similarly divided tribe was brought together through the symbolic resurrection of a revered ancestor and his invocation of the historical memory of the tribe and its connections to important symbolic sites. *A River Runs Through It* is a film that itself alters the political landscape by depicting a Montana that is rapidly disappearing and may soon be irretrievably lost. That vision symbolizes the West as many people want it to remain, and therefore may serve as a catalyst for consensus in conservation policies that would preserve the natural beauty of the region. Persuasion is an essential component of the policy process (Majone, 1989), and skilled alteration of the rhetorical and symbolic situation may be the most advisable technique to persuade the public to favor conservation policies.

If community consensus on water conservation can be developed, enactment of conservationist water policy is likely. Most states in the West statutorily require consideration of the public interest in water allocation, but the standard is vaguely defined (Longo and Elder, 1994) or rarely applied (MacDonnell, 1998). Public interest language in statutes could be used to enact conservation of water resources to benefit ecosystem health, small-scale irrigation farming, and recreation activities like whitewater rafting and fly-fishing. However, public interest language can easily be disregarded if there is no consensus among the public regarding its interests.

When consensus develops in favor of conservation, it has the potential to derail development projects that benefit even the wealthiest and most powerful political forces. A good historical example is the battle over Mono Lake, where local residents successfully used public interest language to derail Los Angeles's plan to divert water from the streams that feed Mono Lake, 250 miles distant from Los Angeles, to supplement the city's water supply (Hart, 1996). Similar cases have occurred in the Great Plains. For example, conservationist citizen groups have successfully used the requirements for critical habitat protection in the Endangered Species Act to stop or delay development of more than a dozen Platte River water projects (Aiken, 1999). Aiken also notes several federal court rulings that prioritize endangered species and critical habitat protection over municipal and industrial water uses (Aiken, 1999). The statutes at issue in those cases require consideration of the public interest. The court rulings support the conclusion that when citizen groups are able to frame water conservation as a broader public interest than municipal or industrial use, they are likely to succeed in blocking development.

184

Substantial public support for conservation is a key to success. Hassler and O'Connor (1986) observe that conservation groups are most successful either as sponsoring parties or *amicus curiae* in U.S. Supreme Court cases when there is a high level of coordination and cooperation across several groups. The strength in numbers and cohesion of citizen-activists help to shape public perceptions to favor conservation. These examples should provide some encouragement for the conservation-minded citizen: if people can alter the rhetorical and symbolic situation to develop consensus for conservation of water resources as the public interest, water can continue to flow downstream, where it belongs.

REFERENCES

Adatto, K. 1993. *Picture Perfect: The Art and Artifice of Public Image-Making.* New York: Basic Books.

Aiken, J. D. 1999. Balancing Endangered Species Protection and Irrigation Water Rights: The Platte River Cooperative Agreement. *Great Plains Natural Resources Journal,* 3, no. 2:119-58.

Arendt, H. 1958. *The Human Condition.* Chicago: University of Chicago Press.

Berry, W. 1977. *The Unsettling of America: Culture and Agriculture.* San Francisco: Sierra Club Books.

Brown, F. L., and H. M. Ingram. 1987. *Water and Poverty in the Southwest.* Tucson: University of Arizona Press.

Hariman, R. 1995. *Political Style: The Artistry of Power.* Chicago: University of Chicago Press.

Hart, J. 1996. *Storm over Mono: The Mono Lake Battle and the California Water Future.* Berkeley: University of California Press.

Hassler, G. L., and K. O'Connor. 1986. Woodsy Witchdoctors versus Judicial Guerillas: The Role and Impact of Competing Interest Groups in Environmental Litigation. *Boston College Environmental Affairs Law Review,* 13 (Summer): 487-520.

Kemmis, D. 1990. *Community and the Politics of Place.* Norman: University of Oklahoma Press.

Longo, P. J., and B. Elder. 1994. Judicial Recognition of the Public Interest in Water Recreation: Nebraska and United States Supreme Court Realities. *Public Land Law Review,* 15: 199-218.

Longo, P. J., and R. D. Miewald. 1989. Institutions in Water Policy: The Case of Nebraska. *Natural Resources Journal,* 29 (Summer): 751-62.

MacDonnell, L. J. 1998. Marketing Water Rights. *Forum for Applied Research and Public Policy,* 13, no. 3: 52-56.

MacIntyre, A. 1984. *After Virtue.* Notre Dame, IN: University of Notre Dame Press.

Maclean, N. (1976) 1989. *A River Runs Through It.* Chicago: University of Chicago/ Pennyroyal Press.

Majone, G. 1989. *Evidence, Argument, and Persuasion in the Policy Process.* New Haven, CT: Yale University Press.

McWilliams, W. C. 1995. Two-Tier Politics and the Problem of Public Policy. In *New Politics of Public Policy,* M. K. Landy and M. A. Levin (eds.), 268-76. Baltimore: Johns Hopkins University Press.

Pollan, M. 1991. *Second Nature: A Gardener's Education.* New York: Dell Publishing.

Reisner, M. 1993. *Cadillac Desert: The American West and its Disappearing Water.* New York: Penguin Books.

Snyder, G. 1990. *The Practice of the Wild.* New York: North Point Press.

Snyder, G. 1995. *A Place in Space: Ethics, Aesthetics, and Watersheds.* Washington, DC: Counterpoint Press

Stegner, W. 1992. *Where the Bluebird Sings to the Lemonade Springs: Living and Writing in the West.* New York: Penguin Books.

Thomas, C. S., ed. 1991. *Politics and Public Policy in the Contemporary American West.* Albuquerque: University of New Mexico Press.

Final Thoughts

Unlike the panorama of the North American Great Plains, this anthology has come to an end. But in no way do we intend to suggest this book has brought us to some sort of safe landing. We want the readers of this book to consider and reconsider the water issues and policies that mark the Great Plains and the interconnectedness with the regions that surround it.

Continued scrutiny of a stressed Great Plains resource, water, benefits from the embrace of a variety of disciplines and ideas. This work represents a multidisciplinary approach in which water issues on the North American Great Plains were considered in a broad context. While more questions than answers were provided, it is evident that the allocation and conservation of water will remain important to the very life of the North American Plains. As the essays suggest, the quality of life for citizens of the North American Plains depends on the ability of policy-makers to consider many variables, especially cultural and environmental ones, over those resulting in their own mere political and economic gain.

By looking at the present and the past, the essays suggest a path for the future, the path of balance. There is an obvious need to balance supply of our water resources with demand. But within that broad framework, we must balance the needs of humankind with the needs of nature. Do we dam rivers to increase storage capacity, balancing that against the needs of the land and wildlife downstream? Furthermore, we must decide how to distribute water between industrial, agricultural, and municipal needs. Should water automatically flow to those who can pay for it, or is there some basic right to clean water that we all have?

In many regions of the Plains, as well as the country and world, the per capita supply of water is decreasing, making it necessary to rely on marginally productive sources. What can we do to protect the supply of water? Citizens of the Great Plains will be assured a better quality of life if the use and preservation of water is tied to the greater concern of life on the Plains. The effort by the citizens to preserve water will mark the character of the Great Plains.

The topics discussed in this collection of essays, for the most part, are not unique to the Great Plains. Change the names of the places and actors, and the issues could be found in Australia, Africa, or the Middle East. What will continue to distinguish the citizens of the Great Plains is their response to the continuing questions about water.

Notes on Contributors

J. David Aiken is professor of agricultural economics at the University of Nebraska–Lincoln. His areas of research and teaching include agricultural law, environmental law, and water law. He received his J.D. from George Washington University, and his B.A. from Hastings College in Nebraska.

John Anderson is associate professor of political science at the University of Nebraska at Kearney. He earned his Ph.D. at Washington State University, and since then has maintained an interest in the role of politics and culture in public policy. He was born and reared in a small town in Wyoming; that background also informs his thinking on the unique nature of life in rural and wild places.

Charles J. Bicak is professor and chair of biology at the University of Nebraska at Kearney. He received his Ph.D. in range ecology from Colorado State University in Fort Collins. His research interests center around stress physiology of both native and crop plant species. Specific among his interests are plant water use efficiency and pollution ecology, addressing both theoretical and applied questions.

Joan Blauwkamp is assistant professor of political science at the University of Nebraska at Kearney. She received her Ph.D. in political science from the University of Iowa in 1997. Her research has focused on environmental rhetorics and politics in popular cultural media and recruitment rhetorics of environmental groups. She teaches political communication and political behavior courses.

Charles R. Britton is professor of economics at the University of Arkansas. His research interests include water resources and commercial bank finance. Dr. Britton's studies have resulted in publication in the *Forum of the AALS, Review of Regional Studies, International Review and Social Science Quarterly, Land Use Policy,* and other journals. He has served on the Governor's Council of Economic Advisors. He has served twice as President of the Association for Arid Land Studies. He also served as Vice President of the Western Social Science Association.

Steven L. Danver is currently adjunct professor of history at Santa Barbara City College, specializing in American Indian history, environmental history, and the history of the American West. He received his B.A. in religious studies from the University of California, Santa Barbara, and his M.A. in historical studies from Graduate Theological Union in Berkeley, California. He is currently pursuing doctoral studies in history at the University of Utah, and is on the editorial staff of *America, History and Life.*

Brian A. Ellison is associate professor of political science at Southwest Missouri State University. His work on natural resources management, intergovernmental relations, and economic development issues has appeared in *American Review of Public Administration, Environmental Management, Policy Studies Journal, Policy Studies Review, Publius: The Journal of Federalism, Society and Natural Resources,* and *State and Local Government Review.*

Richard K. Ford is professor of economics at the University of Arkansas–Little Rock. His research interests include arid land studies and water resource management. He has served on the board of editors for the Forum of the Association of Arid Land Studies.

Charles Fort holds the Paul W. Reynolds and Clarice Kingston Reynolds Endowed Chair in Poetry and is professor of English at the University of Nebraska at Kearney. A MacDowell Fellow, Fort has received national prizes from the Poetry Society of America, Writer's Voice, and the Randall Jarrell Poetry Prize. Fort's works include: The Best American Poetry 2000, The Best of Prose Poem International 2000, *The Prairie Schooner, Nebraska Poet's Calendar 2001,* and *The Town Clock Burning,* reprinted by Carnegie Mellon University Press, under the Classic Contemporary Series.

Peter J. Longo is professor of political science and director of the Honors Program at the University of Nebraska at Kearney. His areas of research and teaching include constitutional politics, environmental policy, and the politics of science. He received his B.A. from Creighton University and his J.D. and Ph.D. from the University of Nebraska–Lincoln.

Conrad Moore is professor of geography at Western Kentucky University in Bowling Green, Kentucky. He has published twelve articles in professional journals dealing with the climate and biogeography of the Great Plains. Most of these contributions have focused on first-hand observations recorded in journals and diaries of overland travelers in the nineteenth century.

David W. Yoskowitz is assistant professor of economics at Texas A&M International University. His interests revolving around water include water markets and allocation mechanisms and the structure of those markets. His work has appeared in the *Natural Resources Journal, Environmental Law Reporter, International Advances in Economic Research,* among others.

Index